AMAZON WEB SERVICES

:

THE COMPLETE GUIDE FROM BEGINNERS TO ADVANCED FOR AMAZON WEB SERVICES

By Richard Derry

© **COPYRIGHT 2019 BY RICHARD DERRY ALL RIGHTS RESERVED**

This document is geared towards providing exact and reliable information in regards to the topic and issue covered. The publication is sold with the idea that the publisher is not required to render accounting, officially permitted, or otherwise, qualified services. If advice is necessary, legal or professional, a practiced individual in the profession should be ordered.

- From a Declaration of Principles which was accepted and approved equally by a Committee of the American Bar Association and a Committee of Publishers and Associations.

In no way is, it legal to reproduce, duplicate, or transmit any part of this document in either electronic means or printed format. Recording of this publication is strictly prohibited and any storage of this document is not allowed unless with written permission from the publisher. All rights reserved.

The information provided herein is stated to be truthful and consistent, in that any liability, in terms of inattention or otherwise, by any usage or abuse of any policies, processes, or directions contained within is the solitary and utter responsibility of the recipient reader. Under no circumstances will any legal responsibility or blame be held against the publisher for any reparation, damages, or monetary loss due to the information herein, either directly or indirectly.

Respective authors own all copyrights not held by the publisher.

The information herein is offered for informational purposes solely and is universal as so. The presentation of the information is without contract or any type of guarantee assurance.

The trademarks that are used are without any consent and the publication of the trademark is without permission or backing by the trademark owner. All trademarks and brands within this book are for clarifying purposes only and are owned by the owners themselves, not affiliated with this document.

DISCLAIMER

The information contained within this eBook is strictly for educational purposes. If you wish to apply ideas contained in this eBook, you are taking full responsibility for your actions.

The author has made every effort to ensure the accuracy of the information within this book was correct at time of publication. The author does not assume and hereby disclaims any liability to any party for any loss, damage, or disruption caused by errors or omissions, whether such errors or omissions result from accident, negligence, or any other cause.

TABLE OF CONTENT

INTRODUCTION 1

CHAPTER 1 6

OVERVIEW OF AMAZON WEB SERVICES (AWS) 6

    EC2: Server configuration and hosting .. 7

    Compute 8

    Amazon S3: Data storage and movement 9

    Databases, data management 11

    Migration, hybrid cloud 11

    Networking 12

    Development tools and application services 12

    Management, monitoring 13

    Security, governance 14

    Big data management, analytics 14

    Artificial intelligence 15

    Mobile development 16

    Messages, notifications 16

    Elastic Load Balancing: Scalable performance 16

    CloudFront: Deliver a better user experience 17

    Elastic Block Store (EBS): Low-latency instance access 18

Getting Started with AWS 22

Other services 23

CHAPTER 2 24

Understanding the Amazon Business Philosophy 24

Measuring the scale of AWS 26

Checking the bottom line........................28

The AWS Infrastructure 28

Making hard hardware decisions............29

Examining Amazon's software infrastructure strategy 31

The AWS Ecosystem..............................34

Counting Up the Network Effects Benefit 37

AWS versus Other Cloud Providers.......40

The rise of shadow IT 42

Getting Ready for the 21st Century 43

Chapter 3..45

Introducing the AWS Management Console 45

Setting Up Your Amazon Web Services Account 46

The Silicon Valley Education Foundation runs on AWS 50

Accessing Your First AWS Service 51

Loading Data into S3 Buckets................54

S3 URL Naming Conventions 56
Last Words on the AWS Management Console 57
Things are not what they seem 58

CHAPTER 4 59

AWS Platform Services 59

Searching with CloudSearch 60
CloudSearch resources 64
    CloudSearch scope 64
    CloudSearch cost 65
PBS runs on AWS (CloudSearch) 66
    Elastic Transcoder scope 70
    Elastic Transcoder cost 71
Simple Queue Service 71
    Simple Queue Service overview 73
    SQS scope 76
    SQS use 78
Simple Notification Service 78
    SNS overview 80
    SNS scope 84
    SNS cost 84
Simple E-Mail Service 85
    SES overview 87
    SES scope 90
Simple Workflow Service 91

    SWF overview 92

    SWF scope 94

    SWF cost... 94

Dealing with Big Data with the Help of Elastic MapReduce..94

    EMR scoping.. 99

    EMR cost 99

CHAPTER 5 101
CLOUD COMPUTING................................101

What is Cloud Computing? 101

Six Advantages of Cloud Computing 102

Types of Cloud Computing 103

    Cloud Computing Models 103

Cloud Computing Deployment Models .105

Global Infrastructure 106

Security and Compliance 107

    Benefits of AWS Security 109

    Compliance 109

AWS Management Console 110

Analytics 111

AWS Lake Formation...............................112

Amazon Managed Streaming for Kafka (MSK) 113

CHAPTER 6 115
Application Integration 115

AWS Step Functions 115
    Amazon MQ 115
    Amazon SQS 116
    Amazon SNS 117
    Amazon SWF 117
AR and VR 117
    Amazon Sumerian 117
AWS Cost Management 118
    AWS Cost Explorer 118
    AWS Budgets 118
    AWS Cost & Usage Report 119
    Reserved Instance (RI) Reporting 119
Blockchain 120
    Amazon Managed Blockchain 120
Business Applications 121
    Alexa for Business 121
    Amazon WorkDocs 121
    Amazon WorkMail 122
    Amazon Chime 122
Database 123
    Amazon Aurora 123
    Amazon RDS 124
    Amazon RDS on VMware 124
    Amazon DynamoDB 125
    Amazon ElastiCache 125

Amazon Neptune 126

Amazon Quantum Ledger Database (QLDB) 127

Amazon Timestream 129

Desktop and App Streaming 130

    Amazon WorkSpaces 130

    Amazon AppStream 2.0 130

RICHARD DERRY

INTRODUCTION

You may be forgiven if you're puzzled about how Amazon, which started out as an online bookstore, has become the leading cloud computing provider. This chapter solves that mystery by discussing the circumstances that led Amazon into the cloud computing services arena and why Amazon Web Services, far from being an oddly different offering from a retailer, is a logical outgrowth of Amazon's business.

A part of this book also compares Amazon's cloud offering to other competitors in the market and explains how its approach differs. As part of this comparison, I present some statistics on the size and growth of Amazon's offering, while describing why it's difficult to get a handle on its exact size.

Another part concludes with a brief discussion about the Amazon Web Services ecosystem and why it is far richer than what Amazon itself provides — and why it offers more value for users of Amazon's cloud service.

But before I reveal all the answers to the Amazon mystery, I answer an even more fundamental question: What is all this cloud computing stuff, anyway?

Cloud Computing Defined

I believe that skill is built on a foundation of knowledge. Anyone who wants to work with Amazon Web Services (AWS, from now on) should have a firm understanding of cloud computing — what it is and what it provides. IaaS, Paas, SaaS As a general overview, cloud computing refers to the delivery of computing services from a remote location over a network. The National Institute of Standards and Technology (NIST), a U.S. government agency, has a definition of cloud computing that is generally considered the gold standard. Rather than trying to create my own definition, I always defer to NIST's definition. The following information is drawn directly from it.

Cloud computing is a model for enabling ubiquitous, convenient, ondemand network access to a shared pool of configurable computing resources (e.g., networks, servers, storage, applications, and services) that can be rapidly provisioned and released with minimal management effort or service provider interaction.

This cloud model is composed of five essential characteristics:

✓ **On-demand self-service:** A consumer can unilaterally provision computing capabilities, such as server time and network storage, automatically as needed without requiring human interaction with each service provider.

✓ **Broad network access**: Capabilities are available over the network and accessed via standard mechanisms that promote use by heterogeneous thin or thick client platforms (such as mobile phones, tablets, laptops, and workstations).

✓ **Resource pooling**: The provider's computing resources are pooled to serve multiple consumers using a multi-tenant model, with different physical and virtual resources dynamically assigned and reassigned according to consumer demand. There's a sense of so-called location independence, in that the customer generally has no control or knowledge over the exact location of the provided resources but may be able to specify location at a higher level of abstraction (by country, state, or data center, for example). Examples of resources are storage, processing, memory, and network bandwidth.

✓ **Rapid elasticity**: Capabilities can be elastically provisioned and released, in some cases automatically, to scale rapidly outward and inward commensurate with demand. To the consumer, the capabilities available for provisioning often appear to be unlimited and can be appropriated in any quantity at any time.

✓ **Measured service**: Cloud systems automatically control and optimize resource use by leveraging a metering capability at a level of abstraction that's appropriate to the type of service (storage, processing, bandwidth, or active user accounts, for example). Resource usage can be monitored, controlled, and reported, providing transparency for both the provider and consumer of the utilized service.

Cloud computing is commonly characterized as providing three types of functionality, referred to IaaS, PaaS, and SaaS, where aaS is shorthand for "as a service" and service implies that the functionality isn't local to the user but rather originates elsewhere (a location in a remote location accessed via a network). The letters I, P, and S in the acronyms refer to different types of functionality, as the following list makes clear:

✓ **Infrastructure as a Service (Iaas):** Offers users the basic building blocks of computing: processing, network connectivity, and storage. (Of course, you also need other capabilities in order to fully support IaaS functionality — such as user accounts, usage tracking, and security.) You would use an IaaS cloud provider if you want to build an application from scratch and need access to fairly low-level functionality within the operating system.

√ **Platform as a Service (PaaS):** Instead of offering low-level functions within the operating system, offers higher-level programming frameworks that a developer interacts with to obtain computing services. For example, rather than open a file and write a collection of bits to it, in a PaaS environment the developer simply calls a function and then provides the function with the collection of bits. The PaaS framework then handles the grunt work, such as opening a file, writing the bits to it, and ensuring that the bits have been successfully received by the file system. The PaaS framework provider takes care of backing up the data and managing the collection of backups, for example, thus relieving the user of having to complete further burdensome administrative tasks.

√ **Software as a Service (SaaS):** Has clambered to an even higher rung on the evolutionary ladder than PaaS. With SaaS, all application functionality is delivered over a network in a pretty package. The user need do nothing more than use the application; the SaaS provider deals with the hassle associated with creating and operating the application, segregating user data, providing security for each user as well as the overall SaaS environment, and handling a myriad of other details.

As with every model, this division into I, P, and S provides a certain explanatory leverage and seeks to make neat and clean an element that in real life can be rather complicated. In the case of IPS, the model is presented as though the types are cleanly defined though they no longer are. Many cloud providers offer services of more than one type. Amazon, in particular, has begun to provide many platform-like services as it has built out its offerings, and has even ventured into a few full-blown application services that you'd associate with SaaS. You could say that Amazon provides all three types of cloud computing.

Private-versus-public cloud computing

If you find the mix of I, P, and S in the preceding section confusing, wait 'til you hear about the whole private-versus-public cloud computing distinction.

Note the sequence of events:

1. Amazon, as the first cloud computing provider, offers public cloud computing — anyone can use it.

2. Many IT organizations, when contemplating this new Amazon Web Services creature, asked why they couldn't create and offer a service like AWS to their own users, hosted in their own data centers. This on premise version became known as private cloud computing.

3. Continuing the trend, several hosting providers thought they could offer their IT customers a segregated part of their data centers and let customers build clouds there. This concept can also be considered private cloud computing because it's dedicated to one user. On the other hand, because the data to and from this private cloud runs over a shared network, is the cloud truly private?

4. Finally, after one bright bulb noted that companies may not choose only public or private, the term hybrid was coined to refer to companies using both private and public cloud environments.

CHAPTER 1

OVERVIEW OF AMAZON WEB SERVICES (AWS)

Amazon Web Services (AWS) is a comprehensive, evolving cloud computing platform provided by Amazon. It provides a mix of infrastructure as a service (IaaS), platform as a service (PaaS) and packaged software as a service (SaaS) offerings.

Amazon Web Services (AWS) is the market leader in IaaS (Infrastructure-as-a-Service) and PaaS (Platform-as-a-Service) for cloud ecosystems, which can be combined to create a scalable cloud application without worrying about delays related to infrastructure provisioning (compute, storage, and network) and management.

With AWS you can select the specific solutions you need, and only pay for exactly what you use, resulting in lower capital expenditure and faster time to value without sacrificing application performance or user experience.

AWS launched in 2006 from the internal infrastructure that Amazon.com built to handle its online retail operations. AWS was one of the first companies to introduce a pay-as-you-go cloud computing model that scales to provide users with compute, storage or throughput as needed.

Amazon Web Services provides services from dozens of data centers spread across availability zones (AZs) in regions across the world. An AZ represents a location that typically contains multiple physical data centers, while a region is a collection of AZs in geographic proximity connected by low-latency network links. An AWS customer can spin up virtual machines (VMs) and replicate data in different AZs to achieve a highly reliable infrastructure that is resistant to failures of individual servers or an entire data center.

New and existing companies can build their digital infrastructure partially or entirely in the cloud with AWS, making the on-premise data center a thing of the past. The AWS cloud ensures infrastructure reliability, compliance with security standards, and the ability to instantly grow or shrink your infrastructure to meet your needs and maximize your budget, all without upfront investment in equipment.

More than 100 services comprise the Amazon Web Services portfolio, including those for compute, databases, infrastructure management, application development and security. These services, by category, include:

EC2: Server configuration and hosting

Deploy your army of servers with Amazon EC2. In just minutes bring virtual machines–called instances–online. Select AMI's with operating system of your choice (Linux or Windows) and start deploying your clusters. There are three popular instance types when creating EC2 instances:

• Compute optimized. Used for instances that will require extremely high request rates, this configuration leverages industry leading processors.

• Memory optimized. These instances are built with the most efficient per-GB memory cost.

• Storage optimized. Storage optimized EC2 instances access extremely fast SSD storage to serve data retrieval requests with lightning speed.

For more information about EC2 instances and the different series that can be leveraged against your specific needs, check out the AWS Instance Types page.

Compute

Amazon Elastic Compute Cloud (EC2) provides virtual servers -- called instances -- for compute capacity. The EC2 service offers dozens of instance types with varying capacities and sizes, tailored to specific workload types and applications, such as memory-intensive and accelerated-computing jobs. AWS also provides an Auto Scaling tool to dynamically scale capacity to maintain instance health and performance.

The Amazon EC2 Container Service and EC2 Container Registry enable customers to work with Docker containers and images on the AWS platform. A developer can also use AWS Lambda for serverless functions that automatically run code for applications and services, as well as AWS Elastic Beanstalk for PaaS. AWS also includes Amazon Lightsail, which provides virtual private servers, and AWS Batch, which processes a series of jobs.

Amazon S3: Data storage and movement

To build a powerful cloud application you need scalable storage. AWS Simple Storage Services (S3) provides all the needed tools to store and move data around the globe using 'buckets.'

AWS S3 simplifies all of your storage needs into containers called buckets, then lets you choose where and how to store them.

There are four ways to designate buckets in AWS and the cost to store them varies greatly.

1) Amazon Standard Storage

For data that are frequently accessed, such as logs for the last 24 hours or a media file that is being accessed frequently, Amazon Simple Storage offers affordable, highly available storage capacity that can grow as quickly as your organization needs. You are charged by the gigabyte used and number of requests to access, delete, list, copy or getting a data in S3. Expensive storage arrays are not required to get a new endeavor off the ground.

2) Amazon Infrequent Access Storage

Using the S3 interface, monitor and manage resources that are necessary for your operation but are used far less frequently. By designating these buckets as infrequent access, data availability will be 99.9% (amounting to less than 9 hours of downtime in a year) as compared to 99.99% availability (less than an hour of downtime in a year) for standard storage but can be stored for far less cost per gigabyte than Standard Storage buckets.

3) Amazon Glacier

AMAZON WEB SERVICES

For deep storage items that must be retained but are rarely used, Amazon Glacier provides long-term archiving solutions. Data stored in Glacier can take hours to retrieve instead of seconds, but the cost is a fraction of standard storage. With redundant data sites all over the world Glacier ensures your archival data is secure and safe no matter what happens.

4) Amazon Reduced Redundancy Storage

This service allows for the storing of non-essential, easily reproducible data, without the same amount of redundancy and durability as their higher-level storage tiers.

Understanding these storage concepts in S3 is essential before building your cloud architecture. You can learn more about the intricacies of S3 here. Now let's take a look at some of the other AWS services that can be used to build scalable cloud application.

Amazon Simple Storage Service (S3) provides scalable object storage for data backup, archival and analytics. An IT professional stores data and files as S3 objects -- which can range up to 5 GB -- inside S3 buckets to keep them organized. A business can save money with S3 through its Infrequent Access storage tier or use Amazon Glacier for long-term cold storage.

Amazon Elastic Block Store provides block-level storage volumes for persistent data storage for use with EC2 instances, while Amazon Elastic File System offers managed cloud-based file storage.

A business can also migrate data to the cloud via storage transport devices, such as AWS Snowball and Snowmobile, or use AWS Storage Gateway to enable on-premises apps to access cloud data.

Databases, data management

AWS provides managed database services through its Amazon Relational Database Service, which includes options for Oracle, SQL Server, PostgreSQL, MySQL, MariaDB and a proprietary high-performance database called Amazon Aurora. AWS offers managed NoSQL databases through Amazon DynamoDB.

An AWS customer can use Amazon ElastiCache and DynamoDB Accelerator as in-memory data caches for real-time applications. Amazon Redshift offers a data warehouse, which makes it easier for data analysts to perform business intelligence tasks.

Migration, hybrid cloud

AWS includes various tools and services designed to help users migrate applications, databases, servers and data onto its public cloud. The AWS Migration Hub provides a location to monitor and manage migrations from on premises to the cloud. Once in the cloud, EC2 Systems Manager helps an IT team configure on-premises servers and AWS instances.

Amazon also has partnerships with several technology vendors that ease hybrid cloud deployments. VMware Cloud on AWS brings software-defined data center technology from VMware to the AWS cloud. Red Hat Enterprise Linux for Amazon EC2 is the product of another partnership, extending Red Hat's operating system to the AWS cloud.

Networking

An Amazon Virtual Private Cloud (VPC) gives an administrator control over a virtual network to use an isolated section of the AWS cloud. AWS automatically provisions new resources within a VPC for extra protection.

Admins can balance network traffic with AWS load balancing tools, including Application Load Balancer and Network Load Balancer. AWS also provides a domain name system called Amazon Route 53 that routes end users to applications.

An IT professional can establish a dedicated connection from an on-premises data center to the AWS cloud via AWS Direct Connect.

Development tools and application services

A developer can take advantage of AWS command-line tools and software development kits (SDKs) to deploy and manage applications and services. The AWS Command Line Interface is Amazon's proprietary code interface. A developer can also use AWS Tools for Powershell to manage cloud services from Windows environments and AWS Serverless Application Model to simulate an AWS environment to test Lambda functions. AWS SDKs are available for a variety of platforms and programming languages, including Java, PHP, Python, Node.js, Ruby, C++, Android and iOS.

Amazon API Gateway enables a development team to create, manage and monitor custom APIs that let applications access data or functionality from back-end services. API Gateway manages thousands of concurrent API calls at once.

AWS also provides a packaged media transcoding service, Amazon Elastic Transcoder, and a service that visualizes workflows for microservices-based applications, AWS Step Functions.

A development team can also create continuous integration and continuous delivery pipelines with services like AWS CodePipeline, AWS CodeBuild, AWS CodeDeploy and AWS CodeStar. A developer can also store code in Git repositories with AWS CodeCommit and evaluate the performance of microservices-based applications with AWS X-Ray.

Management, monitoring

An admin can manage and track cloud resource configuration via AWS Config and AWS Config Rules. Those tools, along with AWS Trusted Advisor, can help an IT team avoid improperly configured and needlessly expensive cloud resource deployments.

AWS provides several automation tools in its portfolio. An admin can automate infrastructure provisioning via AWS CloudFormation templates, and also use AWS OpsWorks and Chef to automate infrastructure and system configurations.

An AWS customer can monitor resource and application health with Amazon CloudWatch and the AWS Personal

Health Dashboard, and also use AWS CloudTrail to retain user activity and application programming interface (API) calls for auditing.

Security, governance

AWS provides a range of services for cloud security, including AWS Identity and Access Management (IAM), which allows admins to define and manage user access to resources. An admin can also create a user directory with Amazon Cloud Directory, or connect cloud resources to an existing Microsoft Active Directory with the AWS Directory Service. Additionally, AWS Organizations enables a business to establish and manage policies for multiple AWS accounts.

The cloud provider has also introduced tools that automatically assess potential security risks. Amazon Inspector analyzes an AWS environment for vulnerabilities that might impact security and compliance. Amazon Macie uses machine learning technology to protect sensitive cloud data.

AWS also includes tools and services that provide software- and hardware-based encryption, protect against DDoS attacks, provision Secure Sockets Layer and Transport Layer Security certificates and filter potentially harmful traffic to web applications.

Big data management, analytics

AWS includes a variety of big data analytics and application services. Amazon Elastic MapReduce offers a Hadoop framework to process large amounts of data, while Amazon Kinesis provides several tools to process and analyze streaming data.

AWS Glue is a service that handles extract, transform and load jobs, while the Amazon Elasticsearch Service enables a team to perform application monitoring, log analysis and other tasks with the open source Elasticsearch tool.

To query data, an analyst can use Amazon Athena for S3, and then visualize data with Amazon QuickSight.

Artificial intelligence

AWS offers a range of AI model development and delivery platforms, as well as packaged AI-based applications. The Amazon AI suite of tools includes Amazon Lex for voice and text chatbot technology, Amazon Polly for text-to-speech translation and Amazon Rekognition for image and facial analysis. AWS also provides technology for developers to build smart apps that rely on machine learning technology and complex algorithms.

With AWS Deep Learning AMIs, developers can create and train custom AI models with clusters of GPUs or compute-optimized instances. AWS also includes deep learning development frameworks for MXNet and TensorFlow.

On the consumer side, AWS technologies power the Alexa Voice Services, and a developer can use the Alexa Skills Kit to build voice-based apps for Echo devices.

Mobile development

The AWS Mobile Hub offers a collection of tools and services for mobile app developers, including the AWS Mobile SDK, which provides code samples and libraries.

A mobile app developer can also use Amazon Cognito to manage user access to mobile apps, as well as Amazon Pinpoint to send push notifications to application end users and then analyze the effectiveness of those communications.

Messages, notifications

AWS messaging services provide core communication for users and applications. Amazon Simple Queue Service is a managed message queue that sends, stores and receives messages between components of distributed applications to ensure that the parts of an application work as intended.

Amazon Simple Notification Service (SNS) enables a business to send pub-sub messages to endpoints, such as end users or services. SNS includes a mobile messaging feature that enables push messaging to mobile devices. Amazon Simple Email Service provides a platform for IT professionals and marketers to send and receive emails.

Elastic Load Balancing: Scalable performance

Amazon includes a powerful, scalable load balancing solution in AWS Elastic Load Balancer (ELB). ELB ensures that client requests are sent to the appropriate servers and avoiding any server hotspots (over-utilizing one server and under utilizing others)

AWS supports two types of load balancing: classic Load balancing and Application Load Balancing.

• Classic Load Balancing, which analyzes basic network and application data and ensure fault tolerance if one of the EC2 instances running web application happens to fail.

• Application Load Balancing, which looks at content request and routes traffic to the appropriate container or microservice based on the Application content information.

As with most AWS services, you only pay for the services that you use. In case of ELB service, you pay for by an hour and by the amount of data processed.

CloudFront: Deliver a better user experience

Amazon Cloudfront is a global content delivery system that leverages Amazon's vast global infrastructure to deliver contents with optimized speed and cost. CloudFront ensures that content is closer to the users and improves the user experience by delivering the content faster by serving the content from the AWS region closer to the end user.

Cloudfront works seamlessly with AWS services. With no minimum usage commitment, experimenting with configurations and instances to find ways to improve performance is extremely easy.

Elastic Block Store (EBS): Low-latency instance access

AWS Elastic Block Storage provides persistent Block-level storage volumes for your EC2 instances with low latency. It also allows your system to access high speed SSD storage and layer your security with Access Control Lists and encryption.

Amazon Route 53: The AWS DNS service

Handle DNS routing with the high-speed, low cost Route 53 service from AWS. Translate machine hosts and named application to IP addresses and back within your VPC and connect resources like web servers, S3 buckets, and elastic load balancers. Route 53 is the network of DNS Servers hosted in various AWS regions all around the world. Using API, developers can easily automate the configuration changes to Route 53.

Cloudwatch: Monitor your AWS environment

Cloudwatch is the native monitoring service for resources and applications running in AWS. Gather logs and monitor metrics for key resources like:

• Amazon EC2 instance

• Amazon DynamoDB tables

• Amazon RDS DB instances

• Custom metrics generated by your applications and services

With Cloudwatch you can achieve full visibility into all of your AWS components.

Optional AWS Support Services

In addition to the essential cloud services, Amazon offers a host of optional products for enhancing and securing your cloud ecosystem. Here's a look at some of the more popular helper services.

The almost limitless possibilities of AWS are wrangled down into manageable control screens, showing you how all of your virtual gears are meshing together.

Lambda: Functions for optimized compute

Pay only for the actual milliseconds of compute time your code require to execute and avoid complexity and management overhead of configuring and managing underlying AWS infrastructure. AWS Lambda abstracts underlying AWS infrastructure and allows developers to focus on running their code.

AWS Config: Infrastructure management

Keep a bird's eye view on your AWS infrastructure and stay secure and compliant with AWS Config. See up to date resource inventory and track any changes to your infrastructure from one convenient management panel.

Elastic Beanstalk: Build and scale great web apps

As the name implies, it's impossible to grow faster than Elastic Beanstalk, the AWS tool for deploying and managing web applications designed in most of the top programming languages. The adjustable autoscale settings allow your apps to grow and contract as needed to avoid latency and wasted resource utilization.

CloudTrail: Monitor and audit activity

Application program interfaces calls (APIs) take place within your environment at a rate that can vary from a few transactions per minute to millions per second. AWS Cloudtrail captures key information about these transactions, including the source IP address, the event time, and more. The data from CloudTrail is especially critical for meeting security standards and complying with internal audits and standards or regulations such as PCI and HIPAA.

Amazon EFS: Manage your files

Quickly and easily create file systems through a simple web interface with Amazon Elastic File System (EFS). EFS grows and shrinks your file storage system as needed so you never run out of space. Mount one file system to multiple EC2 instances to share common data and workload sources, manage access control lists, and more with EFS.

These services add massive flexibility, scalability, and monitoring features that will help your organization master your VPC environment in AWS.

Special Purpose AWS Apps

Amazon also offers some highly specialized tools for getting the most out of your cloud resources. Amazon's comprehensive list of products and services is expansive, but these are some of the most popular specialty apps.

Kinesis: Optimize data flow

A three-pronged tool for fine tuning multimedia data flow, Amazon Kinesis is a platform for loading and analyzing streaming data. It consists of:

Kinesis Firehose, a streaming analytics service, which ingests up to terabytes of data and send it to other AWS services such as S3, Redshift and AWS ES.

Kinesis Analytics, which lets you collect masses of data via simple SQL queries, with no need to bring in developers versed in complex programming languages.

Kinesis Streams, which takes these torrents of data and interactively helps you develop custom applications for processing it. With Kinesis Streams you can:

- Elastically scale your environment to respond to volume.
- Transform terabytes of raw streaming data into interactive dashboards.
- Generate alerts when critical events take place.
- Trigger automated responses to common problems like latency.
- Integrate with other Kinesis elements and AWS to optimize delivery.

VPC Flow: Analyze your traffic

VPC flow logs are detailed records of the IP traffic passing to and from a lone port, a subnet, or your entire VPC environment. Flow logs enable you to get into the deep analytics details about who is going where and when. The information they provide about your network will help you develop architecture and budget plans for ongoing operations and also allows you to conduct network forensics using the VPC flow logs—including what traffic is worth the compute and storage cost of capturing.

DynamoDB: Fast, easy database access

Amazon DynamoDB is a fully managed NoSQL database service in AWS. It works with document and key-value storage models, and its high availability and flexibility make it perfect for gaming, mobile apps, and more. Spin up databases with ease and save on compute expenses with DynamoDB.

Getting Started with AWS

There is much more to the AWS universe than a short guide can detail. But familiarity with the above fundamentals prepares you to expand your organization's infrastructure to the cloud, build an entirely new environment, and master the art of data logging to ensure compliance and security. You can learn more about AWS in the video below, read up on AWS logging best practices, check out the Sumo Logic App for AWS, or sign up for a free trial.

Other services

Amazon Web Services has a range of business productivity SaaS options. The Amazon Chime service enables online video meetings, calls and text-based chats across devices. A business can also take advantage of Amazon WorkDocs, a file storage and sharing service, and Amazon WorkMail, a business email service with calendaring features.

Desktop and streaming application services include Amazon WorkSpaces, a remote desktop-as-a-service platform, and Amazon AppStream, a service that lets a developer stream a desktop application from AWS to an end user's web browser.

AWS also has a variety of services that enable internet of things (IoT) deployments. The AWS IoT service provides a back-end platform to manage IoT devices and data ingestion to other AWS storage and database services. The AWS IoT Button provides hardware for limited IoT functionality, and AWS Greengrass brings AWS compute capabilities to IoT devices.

AWS pricing models and competition

AWS offers a pay-as-you-go model for its cloud services, either on a per-hour or per-second basis. There is also an option to reserve a set amount of compute capacity at a discounted price for customers who prepay in whole, or who sign up for one- or three-year usage commitments.

CHAPTER 2

UNDERSTANDING THE AMAZON BUSINESS PHILOSOPHY

Amazon Web Services was officially revealed to the world on March 13, 2006. On that day, AWS offered the Simple Storage Service, its first service. (As you may imagine, Simple Storage Services was soon shortened to S3.) The idea behind S3 was simple: It could offer the concept of object storage over the web, a setup where anyone could put an object — essentially, any bunch of bytes — into S3. Those bytes may comprise a digital photo or a file backup or a software package or a video or audio recording or a spreadsheet file or — well, you get the idea.

S3 was relatively limited when it first started out. Though objects could, admittedly, be written or read from anywhere, they could be stored in only one region: the United States. Moreover, objects could be no larger than 5 gigabytes — not tiny by any means, but certainly smaller than many files that people may want to store in S3. The actions available for objects were also quite limited: You could write, read, and delete them, and that was it.

In its first six years, S3 has grown in all dimensions. The service is now offered throughout the world in a number of different regions. Objects can now be as large as 5 terabytes. S3 can also offer many more capabilities regarding objects. An object can now have a termination date, for example: You can set a date and time after which an object is no longer available for access. (This capability may be useful if you want to make a video available for viewing for only a certain period, such as the next two weeks.) S3 can now also be used to host websites — in other words, individual pages can be stored as objects, and your domain name (say, www.example.com) can point to S3, which serves up the pages.

S3 did not remain the lone AWS example for long. Just a few months after it was launched, Amazon began offering Simple Queue Service (SQS), which provides a way to pass messages between different programs. SQS can accept or deliver messages within the AWS environment or outside the environment to other programs (your web browser, for example) and can be used to build highly scalable distributed applications.

Later in 2006 came Elastic Compute Cloud (known affectionately as EC2). As the AWS computing service, EC2 offers computing capacity on demand, with immediate availability and no set commitment to length of use. Don't worry if this description of AWS seems overwhelming at first — in the rest of this book, you can find out all about the various pieces of AWS, how they work, and how you can use them to address your computing requirements.

This chapter provides a framework in which to understand the genesis of AWS, with details to follow. The important thing for you to understand is how AWS got started, how big of a change it represents in the way computing is done, and why it's important to your future. The overall pattern of AWS has been to add additional services steadily, and then quickly improve each service over time. AWS is now composed of more than 25 different services, many offered with different capabilities via different configurations or formats. This rich set of services can be mixed and matched to create interesting and unique applications, limited only by your imagination or needs.

So, from one simple service (S3) to more than 25 in just over six years, and throughout the world — and growing and improving all the time! You're probably impressed by how fast all of this has happened. You're not alone. Within the industry, Amazon is regarded with a mixture of awe and envy because of how rapidly it delivers new AWS functionality. If you're interested, you can keep up with changes to AWS via its What's New web page on the AWS site, at http://aws.amazon.com/about-aws/whats-new

This torrid pace of improvement is great news for you because it means that AWS continually presents new things you can do — things you probably couldn't do in the past because the AWS functionality would be too difficult to implement or too expensive to afford even if you could implement it.

Measuring the scale of AWS

Amazon is the pioneer of cloud computing and, because you'd have to have been living under a rock not to have heard about "the cloud," being the pioneer in this area is a big deal. The obvious question is this: If AWS is the big dog in the market and if cloud computing is the hottest thing since sliced bread, how big are we talking about?

That's an interesting question because Amazon reveals little about the extent of its business. Rather than break out AWS revenues, the company lumps them into an Other category in its financial reports.

Nevertheless, we have some clues to its size, based on information from the company itself and on informed speculation by industry pundits.

Amazon itself provides a proxy for the growth of the AWS service. Every so often, it announces how many objects are stored in the S3 service. Given that pace of growth, it's obvious that the business of AWS is booming. Other estimates of the size of the AWS service exist as well. A very clever consultant named Huan Liu examined AWS IP addresses and projected the total number of server racks held by AWS, based on an estimate of how many servers reside in a rack.

If you consider that each server can support a number of virtual machines (the number would vary, of course, according to the size of the virtual machines), AWS could support several million running virtual machines. Amazon publishes a list of public IP addresses; as of May 2013, there are over four million available in AWS. This number is not inconsistent with Liu's estimated number of physical servers; it's also a convenient place to look to track how much AWS is growing.

Checking the bottom line

Though Amazon doesn't announce how many dollars AWS pulls in, that hasn't stopped others from making their own estimates of the size of AWS business — and their estimates make it clear that AWS is a very large business indeed.

Early in 2012, several analysts from Morgan Stanley analyzed the AWS business and judged that the service pulled in $1.19 billion in 2011. (You gotta love the precision that these pundits come up with, eh?) Other analysts from JP Morgan Chase and UBS have calculated that AWS will achieve 2015 revenues of around $2.5 billion.

The bottom line: AWS is big and getting bigger (and better) every day. It really is no exaggeration to say that AWS represents a revolution in computing. People are doing amazing things with it, and this book shows you how you can take advantage of it.

The AWS Infrastructure

If what Amazon is doing with AWS represents a revolution, as I describe in the previous section, how is the company bringing it about? In other words, how is it delivering this amazing service? Throughout this book, I go into the specifics of how the service operates, but for now I outline the general approach that Amazon has taken in building AWS.

First and foremost, Amazon has approached the job in a unique fashion, befitting a company that changed the face of retail. Amazon specializes in a low-margin approach to business, and it carries that perspective into AWS. Unlike almost every other player in the cloud computing market,

Amazon has focused on creating a low-margin, highly efficient offering, and that offering starts with the way Amazon has built out its infrastructure.

Making hard hardware decisions

Unlike most of its competitors, Amazon builds its hardware infrastructure from commodity components. Commodity, in this case, refers to using equipment from lesser-known manufacturers who charge less than their brand name competitors. For components for which commodity offerings aren't available, Amazon (known as a ferocious negotiator) gets rock-bottom prices.

On the hardware side of the AWS offering, Amazon's approach is clear: Buy equipment as cheaply as possible. But wait, you may say, won't the commodity approach result in a less reliable infrastructure? After all, the brand-name hardware providers assert that one benefit of paying premium prices is that you get higher-quality gear. Well ... yes and no. It may be true that premiumpriced equipment (traditionally called enterprise equipment because of the assumption that large enterprises require more reliability and are willing to pay extra to obtain it) is more reliable in an apples-to-apples comparison. That is, an enterprise-grade server lasts longer and suffers fewer outages than its commodity-class counterpart.

The issue, from Amazon's perspective, is how much more reliable the enterprise gear is than the commodity version, and how much that improved reliability is worth. In other words, it needs to know the cost-benefit ratio of enterprise-versus-commodity.

AMAZON WEB SERVICES

Making this evaluation more challenging is a fundamental fact: At the scale on which an Amazon operates (remember that it has nearly half a million servers running in its AWS service), equipment — no matter who provides it — is breaking all the time.

If you're a cloud provider with an infrastructure the size of Amazon's, you have to assume, for every type of hardware you use, an endless round of crashed disk drives, fried motherboards, packet-dropping network switches, and on and on.

Therefore, even if you buy the highest-quality, most expensive gear available, you'll still end up (if you're fortunate enough to grow into a very large cloud computing provider like, say, Amazon) with an unreliable infrastructure. Put another way, at a very large scale, even highly reliable individual components still result in an unreliable overall infrastructure because of the failure of components, as rare as the failure of a specific piece of equipment may be.

The scale at which Amazon operates affects other aspects of its hardware infrastructure as well. Besides components such as servers, networks, and storage, data centers also have power supplies, cooling, generators, and backup batteries. Depending on the specific component, Amazon may have to use custom-designed equipment to operate at the scale required.

Think of AWS hardware infrastructure this way: If you had to design and operate data centers to deal with massive scale and in a way that aligns with a corporate mandate to operate inexpensively, you'd probably end up with a solution much like Amazon's. You'd use commodity computing equipment whenever possible, jawbone prices down when you couldn't obtain commodity offerings, and custom-design equipment to manage your unusually largescale operation.

Examining Amazon's software infrastructure strategy

Because of Amazon's low-margin, highly scaled requirements, you'd probably expect it to have a unique approach to the cloud computing software infrastructure running on top of its hardware environment, right?

You'd be correct.

Amazon has created a unique, highly specialized software environment in order to provide its cloud computing services. I stress the word unique because, at first glance, people often find AWS different and confusing — it is unlike any other computing environment they've previously encountered. After users understand how AWS operates, however, they generally find that its design makes sense and that it's appropriate for what it delivers — and, more important, for how people use the service.

Though Amazon has an unusual approach to its hardware environment, it's in the software infrastructure that its uniqueness truly stands out. Let me give you a quick overview of its features. The software infrastructure is

✓ Based on virtualization: Virtualization — a technology that abstracts software components from dependence on their underlying hardware — lies at the heart of AWS. Being able to create virtual machines, start them, terminate them, and restart them quickly makes the AWS service possible.

As you might expect, Amazon has approached virtualization in a unique fashion. Naturally, it wanted a low-cost way to use virtualization, so it chose the open source Xen Hypervisor as its software foundation. Then it made significant changes to the "vanilla" Xen product so that it could fulfill the requirements of AWS.

The result is that Amazon leverages virtualization, but the virtualization solution it came up with is extended in ways that support vast scale and a plethora of services built atop it.

✓ **Operated as a service:** I know what you're going to say: "Of course it's operated as a service — that's why it's called Amazon Web Services!" That's true, but Amazon had to create a tremendous software infrastructure in order to be able to offer its computing capability as a service.

For example, Amazon had to create a way for users to operate their AWS resources from a distance and with no requirement for local hands-on interaction. And it had to segregate a user's resources from everyone else's resources in a way that ensures security, because no one wants other users to be able to see, access, or change his resources.

Amazon had to provide a set of interfaces — an Application Programming Interface (API) — to allow users to manage every aspect of AWS.

✓ **Designed for flexibility:** Amazon designed AWS to address users like itself — users that need rich computing services available at a moment's notice to support their application needs and constantly changing business conditions.

In other words, just as Amazon can't predict what its computing requirements will be in a year or two, neither can the market for which Amazon built AWS.

In that situation, it makes sense to implement few constraints on the service. Consequently, rather than offer a tightly integrated set of services that provides only a few ways to use them, Amazon provides a highly granular set of services that can be "mixed and matched" by the user to create an application that meets its exact needs.

By designing the service in a highly flexible fashion, Amazon enables its customers to be creative, thereby supporting innovation. Throughout the book, I'll offer examples of some of the interesting things companies are doing with AWS.

Not only are the computing services themselves highly flexible, the conditions of use of AWS are flexible as well. You need nothing more to get started than an e-mail address and a credit card.

✓ **Highly resilient:** If you took the message from earlier in the chapter about the inherent unreliability of hardware to heart, you now recognize that there is no way to implement resiliency via hardware. The obvious alternative is with software, and that is the path Amazon has chosen. Amazon makes AWS highly resilient by implementing resource redundancy — essentially using multiple copies of a resource to ensure that failure of a single resource does not cause the service to fail. For example, if you were to store just one copy of each of your objects within its S3 service, that object may sometimes be unavailable because the disk drive on which it resides has broken down. Instead, AWS keeps multiple copies of an object, ensuring that even if one

— or two! — objects become unavailable because of hardware failure, users can still access the object, thereby improving S3 reliability and durability.

In summary, Amazon has implemented a rich software infrastructure to allow users access to large quantities of computing resources at rock-bottom prices.

The AWS Ecosystem

Thus far, I haven't delved too deeply into the various pieces of the AWS puzzle, but it should be clear (if you're reading this chapter from start to finish) that Amazon offers a number of services to its users. However, AWS hosts a far richer set of services than only the ones it provides. In fact, users can find nearly everything they need within the confines of AWS to create almost any application they may want to implement. These services are available via the AWS ecosystem — the offerings of Amazon partners and third parties that host their offerings on AWS.

So, in addition to the 25+ services AWS itself offers, users can find services that

√ Offer preconfigured virtual machines with software components already installed and configured, to enable quick use

√ Manipulate images

√ Transmit or stream video

√ Integrate applications with one another

√ Monitor application performance

✓ Ensure application security

✓ Operate billing and subscriptions

✓ Manage healthcare claims

✓ Offer real estate for sale

✓ Analyze genomic data

✓ Host websites

✓ Provide customer support

And really, this list barely scratches the surface of what's available within AWS. In a way, AWS is a modern-day bazaar, providing an incredibly rich set of computing capabilities from anyone who chooses to set up shop to anyone who chooses to purchase what's being offered.

On closer inspection, you can see that the AWS ecosystem is made up of three distinct subsystems:

✓ **AWS computing services provided by Amazon:** As noted earlier, Amazon currently provides more than 25 AWS services and is launching more all the time. AWS provides a large range of cloud computing services — you'll be introduced to many of them over the course of this book.

✓ **Computing services provided by third parties that operate on AWS:** These services tend to offer functionality that enables you to build applications of a type that AWS doesn't strictly offer. For example, AWS offers some billing capability to enable users to build applications and charge people to use them, but the AWS service doesn't support many billing use cases — user-specific discounts based on the size of the company, for example. Many companies (and even individuals) offer services complementary to AWS that then allow users to build richer applications more quickly.

✓ **Complete applications offered by third parties that run on AWS:** You can use these services, often referred to as SaaS (Software as a Service), over a network without having to install them on your own hardware. Many, many companies host their applications on AWS, drawn to it for the same reasons that end users are drawn to it: low cost, easy access, and high scalability. An interesting trend within AWS is the increasing move by traditional software vendors to migrate their applications to AWS and provide them as SaaS offerings rather than as applications that users install from a CD or DVD on their own machines.

As you go forward with using AWS, be careful to recognize the differences between these three offerings within the AWS ecosystem, especially Amazon's role (or lack thereof) in all three. Though third-party services or SaaS applications can be incredibly valuable to your computing efforts, Amazon, quite reasonably, offers no support or guarantee about their functionality or performance.

It's up to you to decide whether a given non-AWS service is fit for your needs. Amazon, always working to make it ever easier to locate and integrate third party services into your application, has created the Amazon Marketplace as your go-to place for finding AWS-enabled applications. Moreover, being part of the Marketplace implies an endorsement by AWS, which will make you more confident about using a Marketplace application.

Counting Up the Network Effects Benefit

The reason the AWS ecosystem has become the computing marketplace for all and sundry can be captured in the phrase network effect, which can be thought of as the value derived from a network because other network participants are part of the network. The classic case of a network effect is the telephone: The more people who use telephones, the more value there is to someone getting a telephone — because the larger the number of telephones being used, the easier it is to communicate with a large number of people.

Conversely, if you're the only person in town with a telephone, well, you're going to be pretty lonely — and not very talkative! Said another way, for a service with network effects, the more people who use it, the more attractive it is to potential users, and the more value they receive when they use the service.

From the AWS perspective, the network effect means that, if you're providing a new cloud-based service, it makes sense to offer it where lots of other cloud users are located — someplace like AWS, for example. This network effect benefits AWS greatly, simply because many people, when they start to think about doing something with cloud computing, naturally gravitate to AWS because it's a brand name that they recognize.

However, with respect to AWS, there's an even greater network effect than the fact that lots of people are using it: The technical aspects of AWS play a part as well.

When one service talks to another over the Internet, a certain amount of time passes when the communication between the services travels over the Internet network — even at the speed of light, information traveling long distances takes a certain amount of time. Also, while information is traveling across the Internet, it's constantly being shunted through routers to ensure that it's being sent in the right direction. This combination of network length and device interaction is called latency, a measure of how much of a delay is imposed by network traffic distance.

In concrete terms, if you use a web browser to access data from a website hosted within 50 miles of you, it will likely respond faster than if the same website were hosted 7,000 miles away. To continue this concept, using a service that's located nearby makes your application run faster — always a good thing. So if your service runs on AWS, you'd like any services you depend on to also run on AWS — because the latency affecting your application is much lower than if those services originated somewhere else.

Folks who build services tend to be smart, so they'll notice that their potential customers like the idea of having services nearby. If you're setting up a new service, you'll be attracted to AWS because lots of other services are already located there. And if you're considering using a cloud service, you're likely to choose AWS because the number of services there will make it easier to build your application, from the perspective of service availability and low latency performance.

The network effects associated with AWS give you a rich set of services to leverage as you create applications to run on Amazon's cloud offering. They can work to reduce your workload and speed your application development delivery by relieving you of much of the burden traditionally associated with integrating external software components and services into your application.

Here are some benefits of being able to leverage the network effects of the AWS ecosystem in your application:

✓ **The service is already up and running within AWS**. You don't have to obtain the software, install it, configure it, test it, and then integrate it into your application. Because it's already operational in the AWS environment, you can skip directly to the last step — perform the technical integration.

✓ **The services have a cloud-friendly licensing model**. Vendors have already figured out how to offer their software and charge for it in the AWS environment. Vendors often align with the AWS billing methodology, charging per hour of use or offering a subscription for monthly access. But one thing you don't have to do is approach a vendor that has a large, upfront license fee and negotiate to operate in the AWS environment — it's already taken care of.

✓ **Support is available for the service**. You don't have to figure out why a software component you want to use doesn't work properly in the AWS environment — the vendor takes responsibility for it. In the parlance of the world of support, you have, as the technology industry rather indelicately puts it, a throat to choke.

✓ **Performance improves**. Because the service operates in the same environment that your application runs in, it provides low latency and helps your application perform better.

Before you start thinking about finding a packaged software application to integrate into your application, or about writing your own software component to provide certain functionality, search the Marketplace to see whether one or more applications already provide the necessary functionality.

AWS versus Other Cloud Providers

Nature abhors a vacuum, and markets abhor monopoly providers, so it stands to reason that competitors always enter an attractive market. Cloud computing is no different: There are a plethora of cloud computing providers. Naturally, you'll want to get the lowdown on how AWS measures up. The most important difference between AWS and almost all other cloud providers revolves around what market they target. To understand that aspect, you must understand the basis of the service they offer. Now, AWS grew out of the capabilities that Amazon developed to enable its developers to rapidly create and deploy applications. The service is focused on making developers more productive and, in a word, happier.

By contrast, most other cloud providers have a hosting heritage: Their backgrounds involve supporting infrastructure for IT operations groups responsible for maintaining system uptime. A significant part of the value proposition for hosting providers has traditionally been the high quality of their infrastructures — in other words, the enterprise nature of their servers, networks, storage, and so on.

This heritage carries several implications about enterprise cloud providers:

√ The focus is on the concerns of IT operations rather than on the concerns of developers. Often, this concern translates as, "The service is not easy to use." For example, an enterprise cloud provider may require a discussion with a sales representative before granting access to the service and then impose a back-and-forth manual process as part of the account setup. By contrast, AWS allows anyone with an e-mail address and a credit card access to the service within ten minutes.

✓ The service itself reflects its hosting heritage, with its functionality and use model mirroring how physical servers operate. Often, the only storage an enterprise cloud service provider offers is associated with individual virtual machines — no object storage, such as S3, is offered, because it isn't part of a typical hosting environment.

✓ Enterprise cloud service providers often require a multiyear commitment to resource use with a specific level of computing capacity. Though this strategy makes it easier for a cloud service provider to plan its business, it's much less convenient for users — and it imposes some of the same issues that they're trying to escape from!

✓ The use of enterprise equipment often means higher prices when compared to AWS. I have seen enterprise cloud service providers charge 800 percent more than AWS. Depending on organization requirements and the nature of the application, users may be willing to pay a premium for these providers; on the other hand, higher prices and the long-term commitment that often accompanies the use of an AWS competitor may strike many users as unattractive or even unacceptable.

The rise of shadow IT

Frustration at being unable to get hold of server resources in a timely fashion has led to the phenomenon of shadow IT: developers bypassing IT proper and obtaining resources themselves.

This phenomenon is powerful and growing — at one conference, I heard a CIO state that he had examined the expense reports submitted to him for reimbursement and found more than 50 different AWS accounts being used by his development staff!

Here's something to consider: Shadow IT is a pejorative term, implying stealth and a definite whiff of illicit behavior. On the other hand, someone engaging in shadow IT might, reasonably enough, think of it as "getting the job done" in the face of existing processes that can stretch out to several months the length of time required to obtain resources.

This conflict is unlikely to subside in the near future. Developers relish the freedom and flexibility that AWS provides, though many IT groups are engaged in a fruitless struggle to go back to "the good old days," where they set the rules.

The conflict will ultimately be resolved in favour of developers. The reason is simple: The application is the way businesses gain value from IT, and applications are often directly tied to revenue-generating offerings (say, a mobile phone app that enables users to order goods or services online). Infrastructure, the province of mainstream IT, is then seen as a necessary evil — the plumbing that supports applications.

The advantage held by developers can be seen in cloud computing market share. One technology analyst told me that, by his estimate, AWS represents 75 percent of the market for cloud service providers. I expect pressure to build on enterprise cloud providers to rapidly improve their offerings to include more developer friendly services. Amazon's six-year head start may make it too elusive to overtake.

If you analyze how well Amazon matches up against the NIST definition of cloud computing (discussed at the beginning if this chapter) when compared with its competitors, AWS usually emerges victorious. In part, that's because AWS was the pioneer, and because the first entrant into a market typically gets to define it. There's more to it than that, though.

Amazon's stroke of genius was to put together an innovative offering addressing a market poorly served by traditional IT practices. Though hosting companies typically serve IT operations groups well, the emphasis on enterprise equipment and high uptime availability frustrates developers trying to get access to resources. Stories of waiting weeks or months for servers to be provisioned are rife within the industry. As you might imagine, developers (and the application managers and executives they work for) longed for a different way of doing things — and that's what AWS offers.

Getting Ready for the 21st Century

This chapter provides an overview of Amazon Web Services. It lets you see how AWS has grown from Amazon's own computing needs and infrastructure to now represent Amazon's response to this simple hypothesis: If we need a flexible, cost-effective, and highly scalable infrastructure, a lot of other organizations could probably use one as well."

From that initial insight, Amazon created the computing platform of the 21st century. Targeted at developers, and provided throughout the world, AWS is undergoing explosive growth as more and more people explore how it can enable them to solve problems that were unsolvable by the traditional methods of managing infrastructure.

I hope that you can't wait to jump in to exploring AWS. This book aims to provide you with knowledge you need so that you, too, can leverage the amazing AWS cloud computing offering.

RICHARD DERRY

Chapter 3

INTRODUCING THE AWS MANAGEMENT CONSOLE

Okay, so you're ready to start working with Amazon Web Services (AWS) and cloud computing. But how? Well, it turns out that the services part of Amazon Web Services refers to the fact that all interaction with Amazon's cloud computing service is performed with the help of numerous Application Programming Interface (API) calls over the Internet. These calls are accomplished by either SOAP or REST interfaces carrying data in XML or JSON formats.

Whew! Sounds complicated.

Never fear. Amazon offers its own, web-based interface to enable users to work with AWS. This interface, the AWS Management Console, hides all the complex details of interacting with the AWS API. You interact with the console, and Amazon's program deals with all the complexity under the hood. In fact, many people never interact with AWS except through the Console — it's that powerful. This chapter introduces you to the Console, steps you through setting up your very own AWS account, and even provides your first taste of cloud computing. You get to interact with AWS's S3 storage service, upload a picture of your choice, and then connect to it over the Internet and display it in your browser. How fun is that?

I provide screen shots of the various screens you see during your introduction to the Console so that you know exactly what you should see and do. By the end of this chapter, you'll be ready to interact with AWS and, more importantly, to learn all about AWS's great computing services. Amazon updates the Management Console screens fairly frequently, so the screenshots in this book may look different than what you see displayed on your terminal. Fortunately, it's usually pretty easy to map functionality from one screen version to another, but I wanted to provide a heads-up before you get worried about seeing a display that looks different from what's in the book. The Management Console display changes are a side effect of the rapid evolution and innovation within AWS.

Setting Up Your Amazon Web Services Account

The first thing to do is to create your very own AWS account. In this multistep process, you sign up for the service, provide your billing information, and then confirm your agreement with AWS to create your account. Ready? Let's jump right in:

1. Point your favorite web browser to the main Amazon Web Services

page at http://aws.amazon.com.

You should see a screen.

On the next screen, you're given the opportunity to sign in with an existing AWS account or set up a new account. You're setting up a new account.

Technically, you can also use your existing Amazon retail account if you have one, although I don't recommend that. Think about it — if you share your AWS account and use a retail identifier for it, down the line someone you're sharing your AWS with may end up buying a nice big flatscreen TV on your dime. So my recommendation is that you set up a new AWS account.

3. Make sure that the I Am a New User radio button is selected, fill in an appropriate e-mail address in the given field, and then click the Sign In Using Our Secure Server button.

AWS takes you to a new page, where you're asked to enter your login credentials.

4. Enter a username, your e-mail address (twice, just to be sure), and the password you want to use (again, twice, just to be sure).

5. Click Continue button.

Doing so brings up the Account Information screen, asking for your address and phone number information. You're asked to select the box confirming that you agree to the terms of the AWS customer agreement.

6. Enter the required personal information, confirm your acceptance of the customer agreement, and then click the Create Account and Continue button.

The next page asks you for a credit card number and your billing address information. Amazon has to be sure to get paid, right?

7. Enter the required payment information in the appropriate fields and then Click Continue.

AMAZON WEB SERVICES

The next page you see is a bit curious-looking. Amazon wants to confirm your identity, so it asks for a phone number it can use to call you.

8. Enter your telephone number in the appropriate field and click the Call Me Now button. AWS displays a pin code on the screen and then calls you on the phone number you supplied.

9. Answer the phone and enter the displayed PIN code on the telephone keypad.

The AWS screen updates to look like.

10. Click the Continue button.

You're asked to wait a bit to have your account set up by AWS, but, in my experience, this is no more than two or three minutes. You'll then be sent an e-mail confirming your account setup; you have to click on a link in that e-mail to complete the account signup process.

After setup is complete, you should see a screen that lists all the services you're already signed up for automatically, just by creating your account. Quite an impressive list, eh?

Here are two important points to take away from this initial account setup:

✓ Your account is now set up as a general AWS account. You can use AWS resources anywhere in the AWS system — the US East or either of the two US West regions, Asia Pacific (Tokyo, Singapore, or Australia), South America (Brazil), and Europe (Ireland). Put another way, your account is scoped over the entirety of AWS, but resources are located within a specific region.

✓ You have given AWS a credit card number to pay for the resources you use. In effect, you have an open tab with AWS, so be careful about how much computing resource you consume. For the purposes of this book, you don't have to worry much about costs — your initial sign-up provides a free level of service for a year that should be sufficient for you to perform the steps in this book as well as experiment on your own without breaking your piggy bank.

If you're concerned about overspending on AWS, Amazon's got your back. You can set a billing alert with a specified amount you don't want to go over; if your AWS total use for a month approaches that number, Amazon will send you an alert. You can enable billing alerts by clicking My Account in the Management Console landing page and then clicking on Account Activity on the subsequent page.

That's it. You're all set up in AWS and ready to begin cloud computing. If you're anything like me, you're eager to go for a bit of a spin, just to see how AWS works. So get ready to do one small task with AWS — store and retrieve a photo from the AWS object storage service knows as S3.

You start by going to the AWS home page and placing the cursor over the My Account/Console button in the upper-right corner of the screen. You should see a menu displayed underneath the cursor. Click the top item listed: AWS Management Console. You then see a page that provides access to all the services you're signed up for, including S3. (You may have to enter your password again to access the Management Console from the pull-down menu).

The pull-down menu on the left side of the page allows you to define your AWS Management Console start page. (It's right there under the Set Start Page heading.) The default is the general landing page, although you can choose any one of the specific pages associated with a particular AWS service. For now, leave your start page as is.

The Silicon Valley Education Foundation runs on AWS

Do you have to be a huge company, like Netflix or Amazon, to take advantage of AWS? Not at all. Let me share a case study I was personally involved in: the Silicon Valley Education Foundation (SVEF). Unlike Amazon.com or Netflix, SVEF is a tiny organization — it has fewer than 30 employees. And, unlike Amazon.com or Netflix, SVEF isn't a sophisticated technology user; though its mission focuses on helping students with vital science, technology, engineering, and math (STEM) skills, its staff isn't strong on IT skills.

SVEF engaged the technology consulting firm I ran at the time to evaluate one of its most critical applications, designed to let teachers contribute, share, and improve lesson plans for their classes. SVEF had engaged an outside firm to design and build the application; it evaluated whether the infrastructure on which the application was running was robust enough to avoid downtime and could support large growth in traffic, which SVEF expected as more teachers adopted the application.

Accessing Your First AWS Service

After you're the proud owner of an AWS account, it's time to do something useful. Start by checking out your S3 resources. To do so, click the S3 link on your AWS Management Console start page.

You're taken to a page that lets you manage your S3 resources. If you have sharp eyes, you'll quickly notice that there's nothing listed on the page. So the first thing you have to do is create a storage resource where you can place your first object.

Before I walk you through the step-by-step process of creating a storage resource, though, I want to talk a bit about terminology. You'll notice on the left side of the S3 screen is a button labeled Create Buckets. Now, you may wonder why something that sounds like you'd buy it at a hardware store is prominently displayed in AWS. The answer is simple: AWS refers to all top level identifiers within S3 as buckets, signifying, you may presume, a place to put stuff to store. (The term bucket is, perhaps, your first exposure to AWS's curious nomenclature, but I assure you it won't be the last!)

That the entire application was hosted on a single server in a colocation facility made the answer obvious: The application wasn't protected against hardware failure, had no redundancy, and would face significant challenges if application use scaled significantly.

We performed a study comparing three options:

Install additional hardware to implement redundancy; implement virtualization to abstract the application from specific hardware and make it easy to migrate to new hardware in case of failure; and use Amazon Web Services.

Our conclusion, based on both economics and the shortage of IT skills within the SVEF organization, was that SVEF should move its application to AWS. SVEF would save money running on AWS, compared to its ongoing hosting fee.

It would also suffer no more than ten minutes of downtime if the AWS hardware were to fail. And, finally, if the application required more resources as a result of heavy use, it would be trivially easy to shut down one application instance and start another, larger one.

Based on this recommendation, SVEF moved its application to AWS, where it has run happily ever since and with little downtime. Inspired by the success of moving this application to AWS, SVEF evaluated all the applications it was running, and within six months migrated all of them to cloud environments.

So even if your organization isn't a giant of technology, you can still use AWS and benefit from it.

The first thing to do, therefore, is to create a bucket. Before you run out and do so, however, keep a few AWS conditions in mind:

✓ Bucket names must be unique within the entire AWS system. The names must be unique across all user accounts. So if I have a pet named Star and decide to name one of my buckets Star in his honor, and somebody else has already named one of her buckets Star, well, I'm out of luck. This isn't terribly convenient, but that's the way it is.

✓ Although bucket names are global (unique across the entire AWS system, in other words), buckets themselves are located in a particular region. Let's say you want a bucket to reflect your company's name — Corpname, for example. If you use Corpname to create a bucket, you'll isolate that name to a single region, even if you want to place objects throughout the world in a bucket associated with your company's name. So, a better strategy is to use a common identifier with region-specific information as part of the bucket name; for example, you can use

Corpname-US-East for a bucket in the eastern US region and Corpname- US-West-Oregon for a bucket in the region associated with Oregon.

✓ Use all lowercase letters in creating a bucket. Even though the official S3 naming rules let you use uppercase letters, the S3 Management Console doesn't allow them for buckets created in most AWS regions. If you try to include uppercase letters in the bucket name, the Console returns an error message. Keep in mind that, although AWS is a wonderful service, it does have its quirks. You can always find a way around them, but don't be surprised when you run into things that don't work just the way AWS says they will.

Enough about terminology and naming conventions — it's time to create your first bucket:

1. On the S3 home page, click the Create Bucket button.

Doing so brings up a screen similar to the one you see.

2. Enter a name for your bucket in the Bucket Name field.

Because this is just an experiment, feel free to choose any name you like — and don't worry — if it's a bucket name that's already in use, AWS lets you know.

3. Choose a region from the Region pull-down menu.

Choose the "Oregon" item.

4. Click the Create button.

AWS creates your new bucket and returns you to the S3 page for managing your resources. There, you see something, which now lists the bucket you just created.

Congratulations! You've now done your first bit of cloud computing. Of course, it's not useful yet — your bucket just sits there, like an empty filing cabinet, so put something into it so that you can see how it all hangs together.

Loading Data into S3 Buckets

I suggest that for your first S3 experiment, you upload a picture that you can then retrieve and see displayed in your browser. You start out on the S3 page for managing your resources.

Look for the bucket you just created. Found it? Good!

1. Click to select the bucket you created.

Doing so opens the bucket, and the right side of the screen lists a number of actions you can take within the bucket.

2. Click the Upload button.

The Upload-Select Files dialog box appears.

3. Click the Add Files button.

4. Using the file selector widget that appears, browse your local file system, select a file to upload, and then click Open at the bottom of the widget.

You return to the Upload-Select Files dialog box.

5. Click the Start Upload button in the bottom-right corner of the dialog box.

After a few seconds, your bucket lists the file you just uploaded. If you click on the Properties button on the upper right, you'll see information on the file.

Uploading the file is half the battle. Now all you have to do, via a browser, is access the picture you just uploaded. Before you can do that, however, you need to set permissions on the object to make it available over the Internet to someone other than the owner (that's you, by the way). To do that, follow these steps:

1. In the listing of uploaded files, click to select the file you just uploaded.

2. Click the Properties button in the upper-right corner of the screen.

Doing so brings up a pane filled with all kinds of information about the selected object. You can also access a file's Properties information by right-clicking a selected file and choosing Properties from the menu that appears.

3. Click the arrow next to Permissions.

The Permissions section expands to show the Permissions information. You should see only yourself listed next to

Grantee as someone able to access the file. You need to add a permission so that others can access the file as well.

4. Click the Add More Permissions link.

An additional drop-down menu (labeled Grantee as well) appears below the first menu.

5. Choose Everyone from this second drop-down menu, select the associated Open/Download check box, and then click Save.

The file is now accessible to everyone. To access it, you only have to track down the URL you want to use.

6. Go back to the Properties screen.

Here you will see a panel of information on the object, including its URL.

7. Copy the URL listed in the Link section, create a new tab in your browser, enter the URL you just copied into the address line, and hit Enter.

S3 URL Naming Conventions

If you take a closer look at the URL, you see that it follows an unusual naming convention. The domain is amazonaws.com, but to the left of the domain is s3-us-west-1. You can probably figure out that it represents the region in which you chose to create your bucket. AWS prepends regional information to its domain in order to direct requests to access the object. DNS can then efficiently and reliably locate the overall resource storing an S3 object.

To the right of the domain is the name of the bucket you created. The bucket name is part of every request to S3 and is included in the URL. Following the bucket name in the URL is the filename of the actual object. In the case of my example, it's the not-very clever name Cat Photo. (Note that S3 replaces spaces in a filename with plus signs.)

A bucket can contain only files. It's possible to create folders within a bucket to allow better organization of files. In fact, folders can contain folders themselves, thus allowing S3 to mimic the conventions of computer file systems.

AWS presents the organization of S3 as a set of buckets containing objects or folders, which can contain other folders or objects.

Last Words on the AWS Management Console

You've accomplished a lot in this chapter: You've become acquainted with the AWS Management Console, set up your own AWS account, and even chalked up your first AWS experience by experimenting with the S3 storage service.

I hope that this information has helped you understand how easy it is to get started with AWS. If you followed the step-by-step instructions in this chapter, you probably spent no more than 30 minutes on them, from typing aws.amazon.com in a browser address window to accessing your first cloud computing resource. I also hope this first taste of AWS has whetted your appetite to learn more about it, because I dive next into the full panoply of AWS services.

AMAZON WEB SERVICES

Things are not what they seem

Keep in mind that although the S3 Management Console presents files in a nice, neat folder structure, there is in fact no hierarchical organization of objects within S3 buckets. It's just a collection of objects spread throughout S3, with arbitrarily complex resource names that contain conventions, like the slash (/), as part of the resource name. S3 is referred to as a flat storage system, which means that all objects reside at the top level, with those resource names appearing to reflect hierarchy, but in fact being nothing but an S3 naming convention.

It's hard to wrap your head around this concept, but it provides AWS with enormous flexibility and scalability.

Because no hierarchical organization is truly present, AWS can add storage capacity to a nearly infinite degree and add it without disrupting what's already in the S3 system. This clever arrangement takes some getting used to, given how most people are familiar with file systems reflecting hierarchical organization.

CHAPTER 4

AWS PLATFORM SERVICES

Amazon Web Services (AWS) gives you the core services S3, EC2, and Amazon EBS and then all the additional services such as IAM, ELB, and Route 53. The AWS platform services, however, are the focus of this chapter — they dial up the level of sophistication, by concentrating on these three areas of functionality:

✓ Services that provide additional application functionality: For example, Amazon's Simple Queue Service (SQS) provides functionality to enable asynchronous communication between your application and its users or perhaps another application. I call them extended services.

✓ Additional applications that commoditize traditional software offerings that are important but have typically been expensive and complex: An example is the recently launched Redshift business intelligence application. Many established vendors occupy this space, and their applications have two characteristics in common: They cost a lot, and they're difficult to use. Redshift aims to simplify the building and running of a business intelligence application and to make it much less expensive to operate.

✓ AWS management tools: The AWS API and Management Console are useful for managing specific AWS services (individual EC2 instances, for example), but they don't provide much help in managing an aggregation of AWS resources that make up an application. How can you define and manage the collection of resources that comprise your application? Amazon offers three separate tools: Elastic Beanstalk, CloudFormation and OpsWorks. You should understand the differences between them so that you can select the right one.

AWS platform services offer the same benefits (and generate the same problems) as the core services. Though they provide useful, easy-to-use functionality at a reasonable price, they present the potential for lock-in — the investment of so many resources in a solution that changing course is almost impossible. In fact, the lock-in potential is probably greater for platform services because they're more closely tied to the AWS environment than many of the additional AWS services discussed in Chapter 8. Therefore, you must carefully evaluate whether the benefit you receive by embracing these services outweighs any concerns you may have regarding AWS lock-in.

Searching with CloudSearch

Search is one of the most useful capabilities on the web, and huge businesses have been built on search. (Ever hear of Google?) However, not all searches need to cover the entire Internet, and some searches shouldn't be public.

For example, you may want to make content on your company's website searchable — or limit who can see the results of a search. The challenge for many companies that want to enable search on their websites or other content repositories is that the quality of the typical search tools associated with content management systems is, to put it bluntly, awful. The situation is worse for companies that want to make a content repository — a big collection of documents dropped down into a file system, rather than an actual content management system — searchable. These environments have no search mechanism (no matter how flawed) available.

Traditionally, if you wanted to make a sophisticated search capability available for your content, whether contained in a content management system or plopped down in a disorganized content repository, your options were unappealing:

✓ Buy expensive proprietary search software and use it to organize search capabilities for your content. This option requires a large financial outlay and makes sense only for high-value content.

✓ Download and use open source search software, which is both capable and inexpensive. This option makes search financially viable for content that isn't necessarily high-value but can be made more useful with search capabilities. The downside is that you still have to

✓ Locate hardware on which to install the search software. You may have to purchase equipment to support your search software.

✓ Install and configure the search software on the hardware you obtain. You need to have detailed knowledge of the search software. Most people have little expertise in this arcane area, but it can't be avoided if you want to enable search on your content.

✓ Manage the hardware and software to ensure that your search software remains up and running. If your content

AMAZON WEB SERVICES

repository grows to the point that the indexes associated with it outgrow the hardware you obtained originally, you're back to the same (unappetizing!) buy-hardware-and-configure-the-software routine you started with.

Obviously, this situation is unsatisfactory — and ripe for disruption. Amazon, sensing an opportunity, launched CloudSearch early in 2012. CloudSearch is based on technology that Amazon uses on its own website, which should indicate its capability as well as its scalability.

CloudSearch is based on A9, a search company that Amazon incubated a number of years ago, when it realized that the ability to search — and search accurately — would be important to its business. A9 is used for searching on Amazon and its subsidiaries. Though the original A9 focused on text search and relevant results, the A9 technology team has branched out to image search and social search, a category that relies on user interaction to add context to regular text search.

CloudSearch is capable of searching structured content, such as word processing files, and unstructured content — commonly referred to as free text, or unstructured collections of text-like web pages or forum posts. Using CloudSearch is relatively straightforward, though a bit tricky to understand. The content you want to search has to be indexed (the data within the content is evaluated so that individual words can be located) so that indexes about the word (as well as the documents associated with the word) can be built. For example, if you want to be able to search a large number of documents about zoos, you need to build an index so that in a search for the word elephant, the search software can return every document containing the word elephant.

You upload the data you want to search into a CloudSearch domain, where the given domain name is the name of a searchable documents database. For data uploads, CloudSearch uses SDF (short for Search Data Format). Though CloudSeach can create SDF on the fly for certain types of data, such as PDF and Word files, for others you have to create the SDF documents yourself in order to upload your data. SDF documents can be formatted in either XML or JSON — two common standards for describing data collections. An SDF record is nothing more than a formatted set of key-value items describing the data you want to be able to search on.

After you upload the SDF documents, CloudSearch analyzes them and creates indexes of all the items you've indicated you want to be able to search on. For example, if you create a set of SDF documents outlining all players in a sport for a given year, you may search on the position played or the number of games played in the year. CloudSearch creates indexes on all fields you identify as searchable. Then you can execute searches against your domain on the fields you've identified as searchable. You must also create access policies, which are analogous to EC2 security groups. You define the IP addresses that you want to allow access to CloudSearch, for both search access and domain administrative access.

(Typically, you'd allow all IP addresses to search via CloudSearch because the most common use case is allowing visitors to a website to search information on the website, but you may restrict search access to employees of your company or a small number of partners.)

Though you can execute searches from the AWS management console, the most common search is conducted via the CloudSearch API or the CloudSearch CLI (command-line interface). If you're adding search capabilities to a website, you use the API method to perform searches on your CloudSearch domain.

CloudSearch resources

CloudSearch maintains a high performance level by keeping all indexes you've created within the memory of EC2 instances. Now, the obvious question is exactly how many EC2 instances will the CloudSearch domain require?

This number, however, isn't one that you control; AWS automatically calculates how many instances your search domain requires and their size. CloudSearch supports three instances sizes: Small, Large, and Extra Large. If required, CloudSearch splits your domain indexes across multiple instances in order to retain them in memory and support fast search performance.

CloudSearch scope

CloudSearch is regionally scoped, which affects where you deploy your CloudSearch domain. If the website you're enabling with CloudSearch is in a particular region, there's no fee for network traffic if your CloudSearch domain resides in the same region. Of course, given that CloudSearch is accessible via an AWS API, searches can be executed from anywhere on the Internet, as well as within other AWS regions.

CloudSearch cost

Here are the hourly instance prices:

✓ Small search: $.10 per hour

✓ Large search: $.39 per hour

✓ Extra Large search: $.55 per hour

And here are the data transfer prices per month:

✓ First 10TB: $.12 per gigabyte

✓ Next 40TB: $.09 per gigabyte

✓ Next 100TB: $.07 per gigabyte

The issue of traffic prices may not be significant, because search results return text documents (both XML and JSON are text-based), which do not require much network traffic to send, so your traffic charges will probably not be that high.

You face incidental charges for batch uploads and re-indexing, which shouldn't add significantly to your overall CloudSearch bill. Managing Video Conversions with Elastic Transcoder Elastic Transcoder, a relatively new service (started in 2013), is conceptually quite simple: It converts video files from one format to another.

AMAZON WEB SERVICES

PBS runs on AWS (CloudSearch)

If you've joined the Downton Abbey craze on public television, you know how popular PBS is. PBS presents quality programming on public television stations throughout the United States, and it's renowned as the home of many highbrow British dramatic series. Part of the PBS strategy is to complement successful series with additional material and video content so that it can build more loyal audiences and increase viewership.

PBS offers streaming video on the web; in addition, over the past few years it has begun to offer streaming video to portable devices like mobile phones and tablets. For all the video PBS serves up, it uses AWS. PBS not only uses EC2 and S3 for processing and hosting videos but also leverages CloudFront to distribute video content — up to a petabyte of content every month.

More recently, as part of an initiative to more deeply support mobile devices, PBS has placed additional emphasis on creating mobile applications. Part of this mobile initiative requires better-quality searching to enable users on limited form factors to still be able to find and enjoy targeted video content. As part of this mobile initiative, PBS uses ElasticSearch to enable search. ElasticSearch offers greater scalability and better performance, and it frees PBS personnel from having to manage search software and infrastructure.

Video transcoding is a widely applied computing task. You'd have to have been living under a rock not to have observed that video is everywhere. Though people have used dedicated video-recording devices for more than 40 years, the rise of smartphones (initiated by the launch of the iPhone in 2007) has supercharged the video trend. Spurred on by the iPhone and, more recently, the iPad, video-enabled smartphones and tablets have flooded the market. Amazon even provides a family of tablets branded Kindle Fire. Every one of them is now a video recording device.

The ease of sharing video via video-hosting services has skyrocketed as well; at the time this book was written, 72 hours of video were uploaded to YouTube every minute of the day. Though it may seem now and then that all 72 of those hours feature cats in funny or heartwarming videos, the truth is that video is now a communication medium used by all types of organizations for all kinds of purposes — entertainment, education, documentation, evidence, and a thousand others.

For many video creators, this explosion of video presents an embarrassment of riches — so many output devices are available, each of which has its own preferred format, that making all the required versions of video to support customer preferences is challenging. Thus, transcoding — the conversion of one video format to another — is now a critical capability for video-producing entities. Being able to take a source video and prepare all the versions required for widely used display devices is now critical for organizations that want to leverage the power of visual communication. AWS has been part of the transcoding mix for a long time. In fact, when Netflix first started its video-streaming service, AWS was there as part of its video-transcoding strategy. AWS combined with video transcoding is a natural fit: S3 is a great choice to store the original and transcoded versions of a video, and EC2 can naturally host the compute-intensive transformation process. No statistics are available to indicate what percentage of total AWS workload is represented by video transcoding, but it's probably a significant portion.

The workload associated with transcoding can be erratic — in fact, depending on the organization and the type of videos it creates, transcoding workloads may vary by as much as 1,000 percent over a given timeframe. If your organization is using AWS for transcoding purposes, the service may be perfect from an ease-of-access point of view, but such highly variable workloads impose significant management challenges. Translation? You'll likely need to grow and shrink your EC2 processing pool quite a bit to meet transcoding requirements.

Given these facts, the launch of Elastic Transcoder was a foregone conclusion: It helps organizations perform useful video transcoding in AWS but removes the management overhead. Elastic Transcoder, which is designed to simplify common transcoding tasks, lets you designate videos that need to be transcoded and automatically pulls individual videos from S3 storage, performs the transcoding operation, and then places the transcoded versions into S3 storage. Using Elastic Transcoder, you specify the characteristics of the output format you want for your videos, though it also provides a number of preconfigured popular output formats for iPhone, iPad, and, of course, Kindle Fire. You can operate Elastic Transcoder from the AWS Management Console, but it also offers a RESTful interface so that applications can call the service on their own. The RESTful interface is likely to represent the majority of the service's use because many online video applications will transition to Elastic

Transcoder, given its ease of use. Amazon provides language SDKs (software development kits) in a number of languages such as Python and PHP to reduce the burden on developers; instead of having to call the RESTful service directly.

Every transcoding job submitted to Elastic Transcoder is represented as a JSON object, containing the name of the bucket that holds the file to be transcoded, a set of configurations that you want applied to the file (the output formats you want, for example), and a location in which to place the transcoded video.

Elastic Transcoder operates quite simply:

1. Identify the video(s) you want to convert.

2. Create an Elastic Transcoder pipeline or use an existing pipeline.

A pipeline in this context is simply a service endpoint to which jobs are submitted. An AWS account can have several different pipelines, which allows you to separate and prioritize transcoding tasks, if you want. You can, however, have only one pipeline.

3. Use AWS Identity and Access Management (IAM) to create a role for Elastic Transcoder. This step enables Elastic Transcoder to access your resources (say, video files in S3 buckets) to perform transcoding services. If Elastic Transcoder isn't given appropriate access rights, it cannot access your resources and perform transcoding.

4. (Optional) Create a preset containing the settings that you want Elastic Transcoder to apply during the transcoding process. If you are using an existing pipeline, you can use an existing preset or create a new preset. Amazon provides presets to support popular transcoding operations such as formatting for the iPhone, which can be used instead of creating your own preset.

5. Create a job, which represents the transcoding operation for a specific video. The job is submitted in JSON notation.

When the service was originally launched, each output format required a different job; today, you can request multiple outputs in a single job, which reduces your network transfer costs a bit.

6. (Optional) Configure Elastic Transcoder to use AWS's Simple Notification Service (SNS) to provide you with status updates as the job is executed.

7. After the transcoding job is complete, do something with the output videos stored in S3. You can retrieve the video objects from the S3 buckets in which they've been placed, or you can allow access to them from the bucket (with appropriate Access Control List [ACL] settings to allow access, of course). You can even configure the S3 bucket to serve as a CloudFront origin bucket, which then caches the video at the AWS CloudFront endpoints.

That's all there is to using Elastic Transcoder. Amazon takes care of managing the service, the instances on which the service runs, and the queues (pipelines) associated with submitting jobs to the service. You're only responsible for managing the original video file, interacting with Elastic Transcoder, and doing something with the output video files. In other words, Elastic Transcoder enables you to benefit from the process of transcoding videos without suffering the headache of having to manage its details.

Elastic Transcoder scope

Elastic Transcoder is regionally scoped: An individual pipeline resides in a single region, although the service, because it has a RESTful interface, can use S3 buckets associated with your account in other regions. At the time this book was written, Elastic Transcoder wasn't available in all AWS regions, though you can expect that Amazon will soon make Elastic Transcoder available throughout all AWS regions.

Elastic Transcoder cost

Elastic Transcoder offers quite a simple cost model: AWS charges a fixed price per minute of transcoded video. For standard-definition (SD) video, the cost is around $.015 per minute; for high-definition (HD) video, the cost is around $.030 per minute. The cost is slightly higher in certain regions, butno SD transcoding (as of this writing) costs more than $.018 per minute, nor does HD transcoding cost more than $.036 per minute.

Amazon offers a free tier of Elastic Transcoder use. Every month, the first 20 minutes of SD transcoding, or the first 10 minutes of HD transcoding, is provided for free.

Simple Queue Service

It's time now for my favorite AWS service: Simple Queue Service. (It's a geeky choice, I know — but what can I say?) The queue is an awesome system capability, vastly underused by most application designers — which is unfortunate because you end up with complicated, fragile applications that could be improved if they were integrated with queue services.

Now that you're undoubtedly excited about the queue, what exactly is it? The queue concept is dead-easy to understand: It's a communication mechanism between two processing resources that allows them to collaborate on work without needing to operate in a synchronous manner. This description may seem complicated, but the fact is that you use queues in real-life all the time. Say you need your shirts laundered. You can go to the laundry service, hand over your shirts, wait around for the service to finish washing and pressing them, and then take them home. That's one way to do it, but I think you'll agree that it wastes a tremendous amount of your time. You can refer to this mode of operation as synchronous: You call on the laundry service and then wait for it to be complete.

A different way to get your shirts laundered — and the way this service gets done universally throughout the world — is that you take your shirts to the laundry, drop them off, get a claim ticket along with an estimate of when to return to pick them up, go do other errands (which may include dropping off your shoes at a shoe repair place to get new heels installed), and then return on the estimated readiness date to pick up your nice, fresh, clean shirts.

This second mode is asynchronous. You aren't forced to wait for your shirts to be finished — you just put them into the laundry service's work queue and you get a ticket that is then used to track the job. You return at the given time, having allowed the laundry to do its work while letting you go off and do other (hopefully) productive work.

The queue is the ideal tool for a job that's performed by one service and doesn't require the calling service to wait for the results. Elastic Transcoder, the AWS service I discuss in the earlier section "Managing Video Conversions with Elastic Transcoder," is a good example. Many applications that can use video transcoding don't wait for the transcoding process to complete.

Imagine a community website that allows you to upload a video and then makes it available to visitors in formats that are convenient to them, such as iPhone, iPad, Kindle Fire, or a webpage. If you're running the website, you don't want to force users to wait around while videos are transcoded, do you? Especially because the videos being submitted for other people to view, there's no point in making people wait for the transcoding process to complete. The video can be submitted and placed on the queue to be transcoded, leaving the submitter free to do something else (such as explore the rest of your website).

Many, many processing tasks conform to the asynchronous use pattern; as I hint in my Queue Love confession, there are undoubtedly more potential queue use cases than aren't taken advantage of by application designers, which is too bad.

Simple Queue Service overview

SQS lets you create a queue in AWS and then place and retrieve messages from that queue. However, you can also set permissions on a queue to allow access to it that's broader than your account. Being able to enable a broader population to use your queue is useful when you want to allow outside entities, whether a restricted group (say, partners of your company) or the general public, to access it — particularly, being able to submit tasks to your application while not requiring the submitter to wait until the job is complete.

Naturally enough, Amazon has designed SQS to be extremely robust with very high uptime, which imposes some design constraints that, in turn, affect the way SQS operates. You should understand the operation of SQS to ensure that you aren't taken by surprise by the service's behavior.

SQS allows multiple message submitters and retrievers to share a queue, which is a fancy way to say that you can allow your queue to have multiple processes placing messages on the queue and removing them. You can, for example, operate a number of AWS instances designed to retrieve uploads of videos for Elastic Transcoder, ensuring that no transcoding request is delayed by a large job ahead of it in the queue.

One way that Amazon makes SQS robust is that it implements redundant queues behind the scenes; if one queue fails, another, mirror queue can continue operating until the failed queue is restarted. This strategy ensures that no resource failure can ever make SQS unavailable. However, because messages may be spread across the redundant resources, they may not be delivered in the order they were placed on the queue. Unlike some other queue products, SQS doesn't guarantee first-in, first-out (FIFO) delivery. If a submitter splits a job into several messages, the receiver cannot be sure that they will be retrieved in the proper order.

Though nonguaranteed delivery order isn't a problem for many queue-based applications, those that require an ordered sequence of messages need to create a supra-queue coordination mechanism; a sequence order number that's part of the queue message would be appropriate. A message submitter who places multiple messages that are part of a single overall job may place a sequence order number of one in the first message and a total message number of three in that message, indicating to the reader that it needs to receive three messages to make up the entire submission. The receiving application would read the total message number in the first message, recognize that it needs three messages to receive the complete submission, and continue reading until it had retrieved messages two and three.

Despite the lack of a FIFO mechanism. Amazon guarantees that each message is delivered at least once. Until the message is retrieved, it's retained in the queue, waiting to be read. The potential for messages to be retained until they're read can cause a problem if no reader ever requests a message. If this situation occurs often enough, the queue can become backed up with unread messages — and with enough unread messages, even AWS can get overloaded. Therefore, SQS has a message time-out period that defines how long a message is retained in the queue. The default retention period, set at four days, can be adjusted to meet the requirements of the application. Another SQS characteristic to be aware of is that queue messages remain in the queue until they're deleted — even when they've been read. AWS does this because, even if a message is read, it may not be fully acted on — the reading application may crash or otherwise fail to complete the task associated with the message. To avoid situations in which the queue message is read but not fully acted upon due to resource failure, AWS implements a visibility time-out: While a message is being read, it's locked for a period to ensure that no other reader can access it. However, one key task for a reader is to delete the message when the task associated with the message is complete; if the reader fails to delete the message, another process can — when the visibility time-out expires — read the message again and perform the task associated with the message.

Obviously, redundantly performing work isn't a good idea (generally speaking) and, depending on the application, may even cause problems. Therefore, your reading applications must delete SQS messages after they have completed their tasks. The message size in SQS is restricted to 64KB — for many applications, perhaps not a significant restriction; if the complete task is to place someone's name in a database, 256KB is probably more than ample.

On the other hand, you can easily envision queue-based tasks that can be far larger. In the video transcoding example from earlier in this chapter, almost every video submitted would be far larger than 64KB. The reason for this size restriction is, again, to prevent SQS from being overloaded with requests — too many overlarge messages can cause SQS to choke.

So what can you do to overcome this size restriction? It's straightforward: Rather than place the large data payload (in the example, the large video file) in the message, you can, for example, put it in an S3 bucket. You then place the S3 bucket name in the message and place the message in the queue. The queue reader reads the message and retrieves the video from the S3 bucket based on the information contained in the message. This indirect pointer technique is well-established and commonly used with SQS.

SQS scope

SQS is regionally scoped. Each queue is associated with a particular region; when you create an SQS queue, you define which AWS region will serve as your queue's home. However, because SQS is an AWS-provided service, you don't have to place it in a particular availability zone. In fact, Amazon undoubtedly runs each SQS queue in multiple availability zones to ensure robustness and to prevent failure in the unlikely event of an entire availability zone going offline.

The restriction of an SQS queue to a particular region shouldn't be viewed as a problem; each queue comes with a URL to which users can submit jobs from anywhere. Given that AWS doesn't charge for inbound traffic — and therefore no traffic charge is associated with submitting jobs to an SQS queue — the region restriction of SQS has no significant repercussions.

On the other hand, if you're planning to have an EC2 instance within your own account submit messages to the SQS queue, you want the instances and the queue to reside in the same region. Otherwise, the EC2 instances incur charges for traffic sent interregionally — even if they're at the smaller interregional cost.

SQS cost SQS costs $.50 per million SQS requests, where a request is any kind of SQS API call. That means both message submission and retrieval, as well as setting a queue attribute and the like would incur a charge. Naturally, message submission and retrieval represent the vast preponderance of SQS API calls, so those are the sources of most SQS costs. Up to ten messages (if they're less than the 64KB message limit) can be batched together to count as only one submission.

SQS also offers a free-use tier; you receive up to 1 million SQS requests per month for free.

Don't forget that you also pay for traffic sent out of AWS, starting at $.12 per gigabyte and descending as more traffic is sent. The first gigabyte of outbound traffic is free each month.

SQS use

I hope that my introduction to SQS and my overview of the service piques your interest in using it. Queues are extremely useful, and SQS is useful, robust, and extremely cost-effective. The most daunting challenge for most people in using queues is to think about application design differently. Rather than picture a serial progression of tasks within an application, with each task waiting on a previous task to be completed, you have to consider how to disconnect two tasks, make it possible for them to communicate, and notify one another when work is to be done. Using a queue allows an application to avoid all this waiting around and offer a better user experience — which is important. I encourage you to experiment with SQS to see how you can partition your applications into independent entities that use queues to submit tasks to, and retrieve tasks from, one another. After you get the hang of using queues, you'll start to see lots of opportunities to use them, and you'll likely rethink many of your application design decisions.

Simple Notification Service

As the saying goes, Simple Notification Service (SNS) does what it says on the can: It sends notifications about an event via a convenient mechanism as a way of alerting a person or a computer program that something interesting just happened.

The simplicity of this description belies the power of notifications, however. Consider the case of a system administrator who's responsible for the proper operation of an application within AWS. Clearly, if something stops working properly, she needs to know immediately.

One way to make the administrator aware of application problems is to have her be logged on to AWS at all times, to obsessively check the state of the application every second (which is neither efficient nor fun). Another way is to use a notification: After defining at least one condition that the administrator needs to know about (and, presumably, respond to), you create a mechanism to respond to the condition(s). When the mechanism identifies one of these conditions — errors show up on database reads, for example — it sends a notification to one or more people to evaluate the issue and decide whether to take action.

Notifications can be sent in a variety of ways — via e-mail or SMS messages, for example. Moreover, notifications don't even have to be sent to humans; they can be sent (via e-mail) to a program that takes action whenever it receives a message with a given subject.

The notification is a simple concept but extremely powerful in use. System administrators swear by them (and, occasionally, at them, such as when they receive them at 3 in the morning or while on holiday). You may notice a similarity between SNS and SQS — don't both involve an entity submitting a message to a service that then delivers it? Yes, but with SQS, the entity receiving the message has to take action in order to receive it; with SNS, the service sends the message to the receiving entity automatically, with no action required on the part of the receiving entity. This concept is referred to as pull versus push: With SQS, the receiving entity has to pull the information; whereas with SNS, the information is pushed to the receiving entity.

The distinction between pull and push is useful in situations where the event that the receiving entity is on the lookout for is infrequent but extremely important. You wouldn't want a receiving entity polling a message queue every second for an event that occurs once a month; it would be extremely expensive in terms of processing to pay for constant polling on the off chance that this month's event will happen in the next second. With a notification service, receiving entities (which, again, can be either humans or computer programs) can perform their other work, secure in the knowledge that they'll know almost immediately when the occasional event occurs.

SNS overview

I hope that the foregoing makes clear that notifications are extremely useful. On the other hand, they're a fair amount of work to set up and, if a large volume of notifications are being sent, they can be a lot of work to manage. Sounds like a perfect opportunity for AWS, right? You are correct. SNS operates as an AWS service that you create within your account. After you create the service, you're ready to begin distributing notifications.

You can — and probably will — have multiple notification streams within your notification service. You may have one stream for events and messages from your application to system administrators to alert them to possible problems with your application. You may have another notification stream for your application to send messages to users of your application. You almost certainly wouldn't want to mingle messages for those two very different audiences such that one could read notification messages intended for the other.

SNS allows you to send messages to different audiences (or, indeed, to different individuals within the same audience) by setting up separate topics. A topic, in this context, is a specified stream of notifications that one or more entities can publish. I use the term entities because the topic publisher can be a software component (say, a database that sends notifications whenever certain conditions occur) or a person (someone who logs on to the AWS

Management Console and uses it to send a message, for example). Likewise, the notification recipient can be either a human or a software component. Queue messages — like the messages in Amazon's SQS service — can be retrieved by only one entity; by contrast, notifications such as the ones sent by SNS are sent to any entity that is signed up to receive notifications about a particular topic.

Obviously, one key requirement for (successful) notifications is the ability to control who can send or receive notifications — and just because someone wants to receive notifications doesn't mean he should.

The method that SNS uses to control who (or what) can send or receive notifications on a given topic is the topic policy. As the owner of the topic, you can create policies to control who can sign up to send or receive a topic's Notifications. (The entities that do the sending or receiving are known within SNS as principals.)

This list describes your choices for who receives topic notifications:

✓ Individual: You can identify by policy specific individuals within your account who are permitted to send or receive notifications. SNS is integrated with AWS Identity and

AMAZON WEB SERVICES

Access Management (IAM) to manage the individual identities that are SNS principals.

✓ Accounts: You can identify AWS accounts that can act as principals with respect to a particular topic. The AWS account identifier is used to denote an account that can act as a principal for the SNS topic.

✓ Public: You can allow anyone to act as a principal for your SNS topic. It's probably a bad idea for the public to be able to send notifications, but it could very well make sense to allow anyone who is interested in a given topic to receive notifications. Though my SNS examples thus far focus on technical personnel who may need to receive notifications (such as system administrators), you may like a large audience of individuals who aren't part of your account to be notified of an event. An obvious example is to send an e-mail to all subscribers notifying them of a special offer your company is making available — you would simply publish the notification once, and then every person subscribed would receive a notification.

Speaking of everyone getting a notification, you may ask exactly how notifications can be received. SNS is rich in notification protocol options:

✓ SMS messages sent to a phone number: The phone number has to be registered with SNS, naturally enough, but SNS can send topic notifications via SMS. (Charges for receiving the SMS message apply, of course.)

✓ E-mail sent to an e-mail address: This is a common and popular method of sending notifications. The person who registers to receive the notification receives them via whatever e-mail application she uses.

✓ HTTP/HTTPS: A web application can receive notifications over the general public Internet or via secure HTTP. The assumption is that you have a web-based application receiving traffic directed to a URL that's listening on the appropriate port; when a notification is received, the web application displays the message on a web page. Of course, the web-based application doesn't have to display it; it can do any number of other things with the notification, including forwarding it via another protocol or storing it in a database. A good use case for this notification protocol is system administrators who want an ongoing and constantly updated display of application and system events.

✓ Simple Queue Service (SQS): This hero of the previous section can also be a notification recipient. You may wonder why you would want to use SQS as an SNS recipient. It makes sense if you consider this scenario: You need to be sure that you receive and act on important notifications. If you have an application that must be sure that recipients receive notification of events, how can you be sure that the notifications will occur? After all, e-mail may not reach its destination, SMS (at least in my experience) often seems flaky, and it's not easy to be sure that a web application is running at the right time to receive a notification.

Using SQS ensures that a notification is available and can be acted on no matter what; the queue message remains in the queue (up to the message discard time) for as long as it takes for a second entity to retrieve the message. The use of SQS as the SNS delivery protocol increases SNS robustness.

One important requirement to keep in mind with regard to SNS is that any initial recipient sign-ups have to be confirmed, to prevent malicious SNS sign-ups that can flood the recipient with unwanted notifications or, worse, impose significant costs (such as SMS fees). Each notification protocol requires confirmation from the recipient that they (or it) want to receive notifications on the topic. This strategy can be a bit challenging for nonhuman recipients. For example, a web application still receives an initial confirmation message from SNS and must be able to receive, decipher, and respond to the message before beginning to receive notifications. The application's code first must be able to differentiate between the initial invitation and subsequent notifications and then respond correctly to the invitation. Otherwise, SNS decides that the topic subscription was a mistake and refuses to forward notifications to the application.

SNS scope

SNS is a regionally scoped service. However, SNS operates as an AWS service and is accessible from outside the region; therefore, external programs can use SNS. Each topic, upon creation, is assigned an Amazon Resource Name, or ARN. An entity, either human or application, that wants to publish a notification calls the SNS service with the topic's ARN as one of the arguments. Likewise, notifications can be received outside of AWS; SNS forwards them via the selected protocol to wherever the notification recipient is located.

SNS cost

SNS has probably the most unusual pricing of any AWS service because of the various protocols it supports for notification delivery. The basic service is cost-effective:

$.50 per million SNS API requests and you don't pay for the first million SNS API requests per month. However, the cost of the notifications themselves varies, depending on the protocol:

✓ HTTP/HTTPS ($.06 per 100,000 notifications): The first 100,000 notifications per month are free.

✓ E-mail/e-mail-JSON ($2.00 per 100,000 e-mail/e-mail-JSON requests): The first 1,000 requests per month are free. You would use an e-mail-JSON request to send an e-mail notification to an application rather than to a person; the application parses the JSON text to evaluate the notification and then takes an action in response.

✓ SMS ($.75 per 100 notifications): The first 100 notifications per month are free.

✓ SQS: No charge.

Simple E-Mail Service

Let's face it: E-mail is the hardest-working service on the Internet. Though people complain endlessly about it and continually talk about the up-and comers that will make e-mail obsolete (Facebook, anyone?), e-mail continues to flood the Internet — and it's growing all the time.

E-mail is an extremely effective way to communicate. It excels at transmitting large amounts of data (large as compared to Twitter, for example), and it has the virtue of providing a long-lasting record of communication, making it easy to refer to a communication from the past or to reinitiate a discussion by forwarding a previous e-mail.

AMAZON WEB SERVICES

Beyond its virtues as a personal communication tool, e-mail is an excellent vehicle for business communication. Many businesses use e-mail to send information to their customers for many purposes — to acknowledge an order, track a package, respond to a question, and so on.

When tied to recipient demographics, e-mail can be a powerful marketing tool. You can carry out tightly targeted communication at a fraction of the cost of traditional direct marketing mechanisms, with e-mail delivery almost instantaneous compared to "snail mail" timeframes.

One fly remains in the e-mail ointment, though: managing the e-mail server software. It's finicky, it requires constant configuration and tinkering, and it's difficult to manage when e-mail traffic can fluctuate wildly. Companies can use e-mail services, thereby avoiding the management headache, but the high cost of such services can create other headaches.

So there you have it — a core service, one with high and highly variable loads and one that is difficult to manage and costly to boot. It sounds like a job for AWS. And Amazon has stepped up to tackle the job, providing an AWS-based service that provides enormous scalability at a reasonable price:

Simple Email Service (SES).

SES provides an easy-to-use e-mail service that can support a high volume of e-mail. It probably wouldn't surprise you to learn that SES is based on Amazon's own internally developed e-mail application, because Amazon sends out a ton of e-mail every day. Amazon merely polished up its existing service so that it could be used as part of AWS.

SES overview

SES is straightforward, conceptually. E-mail is a well-established set of standards and protocols, so SES implements and supports established e-mail practices. SES supports the Simple Mail Transfer Protocol (SMTP), a venerable protocol for sending e-mail. You submit your e-mail to SES, using one of the supported integration mechanisms, and it sends the e-mail to the recipient — easy as pie.

Of course, this simple story has a few complications, all related to the seductive usefulness of e-mail. Just as companies have found e-mail to be an incredibly easy way to engage with customers and prospects, so too have malefactors who send endless amounts of spam. The potential for SES to be used to distribute spam is quite high, with these potentially terrible consequences for Amazon:

✓ SES can be perceived as a haven for spam, which can lead to customers not wanting to use SES, or perhaps AWS itself.

✓ In an attempt to limit spam, outside parties, such as ISPs, may refuse to accept e-mail from AWS on behalf of their customers.

✓ If an ISP's refusal to accept e-mail makes SES unusable, honest users of SES would be unfairly penalized for using the same AWS service as spammers.

Obviously, none of these outcomes is acceptable to Amazon, so it has implemented a number of SES requirements to avoid problems. Because of these requirements, getting started and then using SES requires you to deal with these constraints that are important to understand as you prepare to use SES:

1. When you sign up to use SES, you must register the domain from which you will send e-mail (say, example.com). Amazon approves your domain registration in a day or two, so be prepared to work on something else while you're waiting. Amazon refers to this process as verifying your domain.

2. After your domain is verified, the individual e-mail addresses from which you'll send e-mail must be verified as well. SES sets a limit of 1,000 verified addresses, so the service is more appropriate for marketing campaigns and application output than for general corporate e-mail support.

3. When you get started, Amazon places you in an SES sandbox, in which you're limited in what addresses you can send e-mail to — these addresses need to be from within your own domain, which prevents you from immediately spamming someone. During the sandbox period, you're limited to 200 e-mails per day, all of which have to be sent to e-mail addresses that are verified by Amazon.

4. When you have established your trustworthiness, Amazon will move you to the Big Leagues: production SES. Even though you're no longer rationed to the sandbox limits, you're not permitted full, unfettered SES use. As you begin, you're limited in the number of e-mails you can send in any single day, and you're limited in how many you can send in any single second. As Amazon gains more confidence in your use of SES, it raises these limits.

AWS offers four ways to interact with SES and send e-mails:

✓ The AWS Management Console: The console allows you to create and send e-mails. This method, which isn't very efficient, is offered primarily to let you test your SES setup and service.

✓ The SES API: You can write directly to the SES API in order to make web service calls and interact with the SES API interface.

✓ Programming language SMTP modules: SMTP is a venerable protocol — most programming languages have modules or libraries that enable the sending of e-mail via SMTP. Note that the use of a programming language SMTP module requires a special SES username and password (different from the account username and password), which must be requested via the AWS Management Console.

✓ AWS programming language SDKs: Amazon itself offers SDK libraries encapsulating the SES API, which can be used in writing programs to interact with SES.

No matter which interaction method you use, SES dutifully sends off however many e-mails you tell it to. In addition to faithfully sending e-mail, SES collects a number of statistics for you — the number of messages that were delivered, bounced (both temporarily and permanently), or rejected and the number of complaints (e-mail refused by a receiving ISP based on its perception of your e-mail as spam). As for rejected e-mails, before sending your e-mail, SES passes it through content filters designed to weed out spam and content that may be perceived as spam; SES lets you know if any sent message is rejected.

Sending e-mails via a programming module or an AWS SDK is relatively easy. It usually calls for setting some variables (the send-to e-mail address, sent from e-mail address, e-mail body content, and the like), and a call to "send" the e-mail to the SMTP service.

Usually, the most difficult part of sending an e-mail is composing the message body — deciding whether it should be plain text or HTML, or both, and how to format the body so that the recipient finds it interesting enough to open. SES doesn't help you with that decision, although it supports both HTML and plain text e-mail. On the other hand, people tend to futz around with e-mail a lot, to get the formatting correct, and then don't touch the formatting settings for months (or years). This task is where the sandbox comes in handy — it's a place to experiment, to be sure that you nail the appropriate e-mail design.

SES scope

SES is regionally scoped and, like all platform services, is accessible from anywhere on the Internet, so it's quite conceivable to use SES as a standalone service, with e-mails sent from an application residing in your own data center.

SES cost

SES costs $.10 per 1,000 e-mails. If you use EC2 or Elastic Beanstalk, you can send 2,000 e-mails free per day.

Standard outbound network traffic charges apply to SES messages, which are based on total traffic size. If you send humongous e-mails, you'll rack up more of a charge than if you send tiny, one-line e-mails. You're also charged for sending e-mail attachments, at the rate of $.12 per gigabyte.

SES considers an e-mail message to be one message sent to one e-mail address. So if you send one e-mail to 100 different recipients, it counts as 100 e-mail messages.

Simple Workflow Service

Simple Workflow Service (SWF) addresses a common challenge in large, distributed applications: how to coordinate all the work between the components of the application, especially when some of the work carried out by a component may depend on the successful completion of work by another component. SWF is the commercial offering of a service that Amazon implemented for its own, internal operations. SWF is a powerful service, but I would say that the initial letter in the acronym (S for Simple) isn't accurate.

Unlike most AWS services, SWF isn't simple to understand or use. On the other hand, the problems that SWF was designed to address are fiendishly complex and undoubtedly require a complex tool to master them. One traditional way to manage complex workflows is to have a human do it. A person kicks off one task, waits for it to complete, starts a second task, waits for it to be done, and so on. This process has a couple very basic problems:

It's slow, and it's boring. It also doesn't scale well. Another method, used in the past, is to write a custom workflow via scripting or code. That approach definitely addresses the challenges of the previous method, but has its own set of challenges. It supports the workflow it's designed for, but as soon as you want another type of workflow, well, you're out of luck. Or you end up trying to generalize your custom workflow and pretty soon you're working full-time on trying

AMAZON WEB SERVICES

to maintain your simple workflow product rather than on doing any . . . you know, work.

Of course, many commercial workflow engines are available to solve these two problems. Though these engines are quite capable, they commonly carry hefty price tags and, given their esoteric nature, aren't easy to get funded.

SWF addresses this problem with a general workflow functionality that's offered and priced like all other AWS services: Use it when you want, and pay for only what you use. If you have a complex workflow that you need to execute, SWF can be a big help.

SWF overview

SWF is a generalizable workflow coordinator, commonly called a workflow engine. To use it, you create these two elements:

✓ Decider: Defines the tasks that your workflow needs to coordinate

✓ Tasks: Do the work that the decider coordinates

Though SWF needs to run in AWS (after all, it's an AWS service, right?), the tasks aren't limited to running within EC2. They can run anywhere. In fact, they don't even have to run — a task can be a human-powered thing. For example, if you implement a printing workflow, one task can be Review Proof with Client, which is a face-to-face task. After receiving positive feedback from the client, the printer's employee can open a web page and click the Approved button for the Review Proof task, and the remainder of the workflow can proceed in an automated fashion. The workflow need not be a sequential series of tasks, either; it can handle concurrent tasks that are run in parallel. A workflow can also include task dependencies, in which a given task cannot start until one or more previous tasks successfully complete.

An SWF decider can include logic to handle task errors and time-outs, for example, enabling it to handle problems that occur within individual tasks. Naturally, you can write workflows to accept input parameters that control how the workflow executes. You can also incorporate timers, signals, and markers in your workflow to help coordinate tasks.

Though SWF provides an API to interact with the service, Amazon has built a fairly full-featured management capability into the AWS Management Console. I think it's fair to say that it expects most SWF users to manage their workflows via the Management Console. SWF can manage workflows that are arbitrarily complex and that may be quite long-running; therefore, it stores the state of the workflow, which can be accessed from the AWS Management Console or via the API so that you can determine where things stand with a given workflow execution. Completed workflow information is also retained and is available for inspection, although you may prefer to delete retained information because AWS imposes a small charge for retaining completed workflow information.

I must warn you: SWF isn't for the faint of heart. However, the SWF section of the AWS Management Console does have a simple application example that demonstrates the power of SWF. This image processing application accepts an input image and converts it to sepia or gray-tone, depending on input it receives via a dialog box. To see SWF in action, check out the example.

SWF scope

SWF is regionally scoped, although it can access AWS resources in other regions as well as non-AWS resources.

SWF cost

AWS imposes several types of charges for SWF, although the aggregate cost is extremely low, unless you execute vast numbers of workflows. For every executed workflow, AWS charges $0.0001. However, you receive 1,000 free workflows per month. If a workflow remains open beyond 24 hours,

AWS imposes a $0.000005 fee per day. If a workflow is retained beyond completion, AWS charges the same $0.000005 per day. AWS provides 30,000 open or retained workdays for free.

AWS also imposes a fee for individual tasks, markers, timers, and signals — $0.000025 per task, signal, timer, or marker. AWS provides 10,000 of these items for free per month. These costs vary slightly by region, but not significantly.

Dealing with Big Data with the Help of Elastic MapReduce

You have to have been living under a rock not to have heard of the term big data. It's a deceptively simple term for an unnervingly difficult problem: how to make sense of the torrents of data flooding into today's applications.

Let me quote a couple factoids to outline the dimension of the big-data challenge. In 2010, Google's chairman, Eric Schmidt, noted that humans now create as much information in two days as all of humanity had created up to the year 2003. Moreover, the research firm IDC projects that the digital universe will reach 40 zettabytes (ZB) by 2020, resulting in a 50-fold growth from the beginning of 2010. In other words, there's lots and lots of data, and its growth is accelerating.

The challenge that big data presents is that most of the established data analytics tools can't scale to manage datasets of the size that many companies want to analyze. For one, traditional business intelligence or data warehousing tools (the terms are used so interchangeably that they're often referred to as BI/DW) are extremely expensive; when applied to very large datasets, you soon face national-debt-type numbers.

Humor aside, the established BI/DW tools have a more serious scalability shortcoming: They're architected with a central analytics engine that reads data from disks, performs analysis, and spits out results. Today, data sizes are so huge that simply sending the data to be analyzed across the network takes too long to perform any useful work. By the time the data is transferred, the insights that can be gleaned from it are obsolete.

Clearly, a new BI/DW analytics architecture and problem approach was called for, and for inspiration the industry reached out to Google. Google has implemented a different approach to gathering data. Its architecture, MapReduce, is based on this simple insight: With so much data, it makes sense to move the processing to the data rather than attempt to move the data to the processing. MapReduce takes a very large datastore that may be spread across hundreds or thousands of machines and formats the data to structure it for the type of analysis you want to perform (that is, it maps the data into an analyzable format), and then you filter the data (reduce the mapped data, in other words) to isolate the information you want to examine.

Google treats its MapReduce implementation as proprietary, but, based on a paper it published, one person implemented an open source version of MapReduce called Hadoop. It's no exaggeration to say that Hadoop has revolutionized the BI/DW industry. In fact, an entire ecosystem of complementary products exists to make Hadoop even more useful.

You've probably already cut to the chase and recognized a familiar refrain:

Hadoop is useful, but complex to install, configure, and manage. Gee, wouldn't it be useful if someone created an easy-to-use, cost-effective Hadoop solution that integrates with the existing ecosystem, allowing established tools that complement Hadoop to be used with this service?

Yes, it would, and Amazon calls its Hadoop solution Elastic MapReduce (EMR). The concept is straightforward:

1. Identify the data source you want to analyze.

This is data located in S3. EMR can handle petabytes (a petabyte is 1,000 terabytes) of data with no problem.

2. Tell EMR how many instances (and of what type) you want the EMR pool to contain.

EMR can use EC2 standard instances or one of the more exotic types, such as High-IO or High-CPU. Each instance offers a certain amount of disk storage for running the Hadoop Distributed File System (HDFS).

The total amount of data you want to analyze dictates the number of instances you require.

3. Set up an EMR job flow.

A job flow can be either of two types:

• Streaming: Programming language mappers and reducers are introduced into EMR and processed across EC2 instances and the data they include.

• Query-oriented: A higher-level data warehouse tool, such as Hive (which provides a Structured Query Language-like interface) can be used to run interactive queries against the data. The output of either type can be stored in S3 and then used for further analysis without requiring an active job flow.

4. Continue running the job flow, running MapReduce programs or higherlevel query languages against the data, until you're finished using the job flow. A job flow can be terminated, which terminates all instances that make up the EMR pool.

Amazon manages the instances within the EMR pool. If an instance terminates unexpectedly, Amazon starts a new instance and ensures that it has the correct data on it to replace the terminated instance. And, of course, Amazon takes care of starting the EMR pool, connecting the instances to one another, and running MapReduce programs or providing higher-level tools for you to use for analysis.

EMR supports these programming languages: Java, Ruby, Perl, Python, PHP, R, Bash, and C++. With respect to these

AMAZON WEB SERVICES

higher-level tools, Amazon provides a wide variety. In addition to Hive (as just mentioned), Amazon also offers Pig (a specialized Hadoop language). Finally, if you want, you can use EMR to output data that can then be imported into a specialized analytics tool like (the curiously named) R.

EMR is one service in which Amazon's pay-only-for-what-you-use philosophy may not be optimal, because transferring and formatting very large datasets to the EMR EC2 instances may take a long time. When you end a job flow, the instances on which the EMR pool is running are terminated and the data discarded. The next time you want to run an analysis, you have to rebuild the EMR pool. So you need to establish a trade-off, to balance the cost of keeping your EMR pool up and running versus the cost of rebuilding it. Clearly, if you plan to run multiple analyses over time against a data pool, it probably makes sense to keep your job flow active.

One interesting characteristic of EMR is that it differs from the other platform services I've already described. The others are "helper" services — useful services that help you build better applications more quickly. By contrast, EMR represents a stand-alone application that's not intended to support an application that the user is writing. Another example of this type of "nonhelper" stand-alone application is Redshift, covered next. I expect that you'll see more of these stand-alone applications, for these reasons:

✓ Its serious reputation: Amazon feels that AWS is now accepted as a serious IT player, and IT is willing to trust it with important use cases. The company is now ready to branch out into areas that provide more direct user benefit in addition to its established infrastructure components that enable users to build their own applications.

✓ The opportunity to expand: Amazon perceives many application domains as ripe for automation and commoditization. As it provides offerings in these domains, its users increasingly benefit, and AWS can become more useful to them, thereby cementing its place as a critical part of their IT environments.

✓ Strategic pricing strategies: AWS recognizes that the high price of current offerings in these application domains prevents many potential users from taking advantage of them; its offerings democratize access to these domains. I'll let you decide whether Amazon is acting purely altruistically in this regard, or perhaps with an element of self-interest.

EMR scoping

EMR is regionally scoped. You should locate your EMR use in the same region where your data resides, if you want to avoid data transfer fees. (Given the kind of data volumes that EMR supports, avoiding these fees can be a big deal.)

EMR cost

The primary cost of EMR is the cost of the EC2 instances on which your EMR pool runs, as well as the S3 storage for your input data and results (assuming, reasonably, that you output results to S3).

In addition, you pay an additional EMR fee per instance. Think of it as an instance surcharge that Amazon imposes to manage the EMR service, install and configure the EMR software on the instances within your EMR pool, and transfer data between all the instances and S3. The EMR surcharge is approximately 25 percent of the instance cost, making it (in my opinion, at least) a modest cost for such a powerful application, compared to the cost of managing Hadoop on your own.

CHAPTER 5

CLOUD COMPUTING

What is Cloud Computing?

Cloud computing is the on-demand delivery of compute power, database storage, applications, and other IT resources through a cloud services platform via the Internet with pay-as-you-go pricing. Whether you are running applications that share photos to millions of mobile users or you're supporting the critical operations of your business, a cloud services platform provides rapid access to flexible and low-cost IT resources. With cloud computing, you don't need to make large upfront investments in hardware and spend a lot of time on the heavy lifting of managing that hardware. Instead, you can provision exactly the right type and size of computing resources you need to power your newest bright idea or operate your IT department. You can access as many resources as you need, almost instantly, and only pay for what you use.

Cloud computing provides a simple way to access servers, storage, databases and a broad set of application services over the Internet. A cloud services platform, such as Amazon Web Services, owns and maintains the network-connected hardware required for these application services, while you provision and use what you need via a web application.

Six Advantages of Cloud Computing

• • Trade capital expense for variable expense – Instead of having to invest heavily in data centers and servers before you know how you're going to use them, you can pay only when you consume computing resources, and pay only for how much you consume.

• • Benefit from massive economies of scale – By using cloud computing, you can achieve a lower variable cost than you can get on your own. Because usage from hundreds of thousands of customers is aggregated in the cloud, providers such as AWS can achieve higher economies of scale, which translates into lower pay as-you-go prices.

• • Stop guessing capacity – Eliminate guessing on your infrastructure capacity needs. When you make a capacity decision prior to deploying an application, you often end up either sitting on expensive idle resources or dealing with limited capacity. With cloud computing, these problems go away. You can access as much or as little capacity as you need, and scale up and down as required with only a few minutes' notice.

• • Increase speed and agility – In a cloud computing environment, new IT resources are only a click away, which means that you reduce the time to make those resources available to your developers from weeks to just minutes. This results in a dramatic increase in agility for the organization, since the cost and time it takes to experiment and develop is significantly lower.

• • Stop spending money running and maintaining data centers – Focus on projects that differentiate your business, not the infrastructure. Cloud computing lets you focus on your own customers, rather than on the heavy lifting of racking, stacking, and powering servers.

• • Go global in minutes – Easily deploy your application in multiple regions around the world with just a few clicks. This means you can provide lower latency and a better experience for your customers at minimal cost.

Types of Cloud Computing

Cloud computing provides developers and IT departments with the ability to focus on what matters most and avoid undifferentiated work such as procurement, maintenance, and capacity planning. As cloud computing has grown in popularity, several different models and deployment strategies have emerged to help meet specific needs of different users. Each type of cloud service and deployment method provides you with different levels of control, flexibility, and management. Understanding the differences between Infrastructure as a Service, Platform as a Service, and Software as a Service, as well as what deployment strategies you can use, can help you decide what set of services is right for your needs.

Cloud Computing Models
Infrastructure as a Service (IaaS)

Infrastructure as a Service (IaaS) contains the basic building blocks for cloud IT and typically provide access to networking features, computers (virtual or on dedicated hardware), and data storage space. IaaS provides you with the highest level of flexibility and management control over your IT resources and is most similar to existing IT resources that many IT departments and developers are familiar with today.

Platform as a Service (PaaS)

Platform as a Service (PaaS) removes the need for your organization to manage the underlying infrastructure (usually hardware and operating systems) and allows you to focus on the deployment and management of your applications. This helps you be more efficient as you don't need to worry about resource procurement, capacity planning, software maintenance, patching, or any of the other undifferentiated heavy lifting involved in running your application.

Software as a Service (SaaS)

Software as a Service (SaaS) provides you with a completed product that is run and managed by the service provider. In most cases, people referring to Software as a Service are referring to end-user applications. With a SaaS offering you do not have to think about how the service is maintained or how the underlying infrastructure is managed; you only need to think about how you will use that particular piece of software. A common example of a SaaS application is web-based email which you can use to send and receive email without having to manage feature additions to the email product or maintain the servers and operating systems that the email program is running on.

RICHARD DERRY

Cloud Computing Deployment Models
Cloud

A cloud-based application is fully deployed in the cloud and all parts of the application run in the cloud. Applications in the cloud have either been created in the cloud or have been migrated from an existing infrastructure to take advantage of the benefits of cloud computing. Cloud-based applications can be built on low-level infrastructure pieces or can use higher level services that provide abstraction from the management, architecting, and scaling requirements of core infrastructure.

Hybrid

A hybrid deployment is a way to connect infrastructure and applications between cloud-based resources and existing resources that are not located in the cloud. The most common method of hybrid deployment is between the cloud and existing on-premises infrastructure to extend, and grow, an organization's infrastructure into the cloud while connecting cloud resources to the internal system. For more information on how AWS can help you with your hybrid deployment, please visit our hybrid page.

On-premises

The deployment of resources on-premises, using virtualization and resource management tools, is sometimes called the "private cloud." On-premises deployment doesn't provide many of the benefits of cloud computing but is sometimes sought for its ability to provide dedicated resources. In most cases this deployment model is the same as legacy IT infrastructure while using application management and virtualization technologies to try and increase resource utilization.

Global Infrastructure

AWS serves over a million active customers in more than 190 countries. We are steadily expanding global infrastructure to help our customers achieve lower latency and higher throughput, and to ensure that their data resides only in the AWS Region they specify. As our customers grow their businesses, AWS will continue to provide infrastructure that meets their global requirements.

The AWS Cloud infrastructure is built around AWS Regions and Availability Zones. An AWS Region is a physical location in the world where we have multiple Availability Zones. Availability Zones consist of one or more discrete data centers, each with redundant power, networking, and connectivity, housed in separate facilities. These Availability Zones offer you the ability to operate production applications and databases that are more highly available, fault tolerant, and scalable than would be possible from a single data center. The AWS Cloud operates in over 60 Availability Zones within over 20 geographic Regions around the world, with announced plans for more Availability Zones and Regions. For more information on the AWS Cloud Availability Zones and AWS Regions, see AWS Global Infrastructure.

Each Amazon Region is designed to be completely isolated from the other Amazon Regions. This achieves the greatest possible fault tolerance and stability. Each Availability Zone is isolated, but the Availability Zones in a Region are connected through low-latency links. AWS provides you with the flexibility to place instances and store data within multiple geographic regions as well as across multiple Availability Zones within each AWS Region. Each Availability Zone is designed as an independent failure zone. This means that Availability Zones are physically separated within a typical metropolitan region and are located in lower risk flood plains (specific flood zone categorization varies by AWS Region). In addition to discrete uninterruptable power supply (UPS) and onsite backup generation facilities, they are each fed via different grids from independent utilities to further reduce single points of failure. Availability Zones are all redundantly connected to multiple tier-1 transit providers.

Security and Compliance
Security

Cloud security at AWS is the highest priority. As an AWS customer, you will benefit from a data center and network architecture built to meet the requirements of the most security-sensitive organizations. Security in the cloud is much like security in your on-premises data centers—only without the costs of maintaining facilities and hardware. In the cloud, you don't have to manage physical servers or storage devices. Instead, you use software-based security tools to monitor and protect the flow of information into and of out of your cloud resources.

An advantage of the AWS Cloud is that it allows you to scale and innovate, while maintaining a secure environment and paying only for the services you use. This means that you can have the security you need at a lower cost than in an on-premises environment.

As an AWS customer you inherit all the best practices of AWS policies, architecture, and operational processes built to satisfy the requirements of our most security-sensitive customers. Get the flexibility and agility you need in security controls.

The AWS Cloud enables a shared responsibility model. While AWS manages security of the cloud, you are responsible for security in the cloud. This means that you retain control of the security you choose to implement to protect your own content, platform, applications, systems, and networks no differently than you would in an on-site data center.

AWS provides you with guidance and expertise through online resources, personnel, and partners. AWS provides you with advisories for current issues, plus you have the opportunity to work with AWS when you encounter security issues.

You get access to hundreds of tools and features to help you to meet your security objectives. AWS provides security-specific tools and features across network security, configuration management, access control, and data encryption.

Finally, AWS environments are continuously audited, with certifications from accreditation bodies across geographies and verticals. In the AWS environment, you can take advantage of automated tools for asset inventory and privileged access reporting.

Benefits of AWS Security

• • Keep Your Data Safe: The AWS infrastructure puts strong safeguards in place to help protect your privacy. All data is stored in highly secure AWS data centers.

• • Meet Compliance Requirements: AWS manages dozens of compliance programs in its infrastructure. This means that segments of your compliance have already been completed.

• • Save Money: Cut costs by using AWS data centers. Maintain the highest standard of security without having to manage your own facility

• • Scale Quickly: Security scales with your AWS Cloud usage. No matter the size of your business, the AWS infrastructure is designed to keep your data safe.

Compliance

AWS Cloud Compliance enables you to understand the robust controls in place at AWS to maintain security and data protection in the cloud. As systems are built on top of AWS Cloud infrastructure, compliance responsibilities will be shared. By tying together governance-focused, audit-friendly service features with applicable compliance or audit standards, AWS Compliance enablers build on traditional programs. This helps customers to establish and operate in an AWS security control environment.

The IT infrastructure that AWS provides to its customers is designed and managed in alignment with best security practices and a variety of IT security standards. The following is a partial list of assurance programs with which AWS complies:

AMAZON WEB SERVICES

- • SOC 1/ISAE 3402, SOC 2, SOC 3
- • FISMA, DIACAP, and FedRAMP
- • PCI DSS Level 1
- • ISO 9001, ISO 27001, ISO 27017, ISO 27018

AWS provides customers a wide range of information on its IT control environment in whitepapers, reports, certifications, accreditations, and other third-party attestations. More information is available in the Risk and Compliance whitepaper and the AWS Security Center.

Amazon Web Services Cloud Platform

AWS consists of many cloud services that you can use in combinations tailored to your business or organizational needs. This section introduces the major AWS services by category. To access the services, you can use the AWS Management Console, the Command Line Interface, or Software Development Kits (SDKs).

AWS Management Console

Access and manage Amazon Web Services through the AWS Management Console, a simple and intuitive user interface. You can also use the AWS Console Mobile Application to quickly view resources on the go.

AWS Command Line Interface

The AWS Command Line Interface (CLI) is a unified tool to manage your AWS services. With just one tool to download and configure, you can control multiple AWS services from the command line and automate them through scripts.

Software Development Kits

Our Software Development Kits (SDKs) simplify using AWS services in your applications with an Application Program Interface (API) tailored to your programming language or platform.

Analytics
Amazon Athena

Amazon Athena is an interactive query service that makes it easy to analyze data in Amazon S3 using standard SQL. Athena is serverless, so there is no infrastructure to manage, and you pay only for the queries that you run.

Athena is easy to use. Simply point to your data in Amazon S3, define the schema, and start querying using standard SQL. Most results are delivered within seconds. With Athena, there's no need for complex extract, transform, and load (ETL) jobs to prepare your data for analysis. This makes it easy for anyone with SQL skills to quickly analyze large-scale datasets.

Athena is out-of-the-box integrated with AWS Glue Data Catalog, allowing you to create a unified metadata repository across various services, crawl data sources to discover schemas and populate your Catalog with new and modified table and partition definitions, and maintain schema versioning. You can also use Glue's fully-managed ETL capabilities to transform data or convert it into columnar formats to optimize cost and improve performance.

AWS Lake Formation

AWS Lake Formation is a service that makes it easy to set up a secure data lake in days. A data lake is a centralized, curated, and secured repository that stores all your data, both in its original form and prepared for analysis. A data lake enables you to break down data silos and combine different types of analytics to gain insights and guide better business decisions.

However, setting up and managing data lakes today involves a lot of manual, complicated, and time-consuming tasks. This work includes loading data from diverse sources, monitoring those data flows, setting up partitions, turning on encryption and managing keys, defining transformation jobs and monitoring their operation, re-organizing data into a columnar format, configuring access control settings, deduplicating redundant data, matching linked records, granting access to data sets, and auditing access over time.

Creating a data lake with Lake Formation is as simple as defining where your data resides and what data access and security policies you want to apply. Lake Formation then collects and catalogs data from databases and object storage, moves the data into your new Amazon S3 data lake, cleans and classifies data using machine learning algorithms, and secures access to your sensitive data. Your users can then access a centralized catalog of data which describes available data sets and their appropriate usage. Your users then leverage these data sets with their choice of analytics and machine learning services, like Amazon EMR for Apache Spark, Amazon Redshift, Amazon Athena, Amazon SageMaker, and Amazon QuickSight.

Amazon Managed Streaming for Kafka (MSK)

Amazon Managed Streaming for Kafka (Amazon MSK) is a fully managed service that makes it easy for you to build and run applications that use Apache Kafka to process streaming data. Apache Kafka is an open-source platform for building real-time streaming data pipelines and applications. With Amazon MSK, you can use Apache Kafka APIs to populate data lakes, stream changes to and from databases, and power machine learning and analytics applications.

Apache Kafka clusters are challenging to setup, scale, and manage in production. When you run Apache Kafka on your own, you need to provision servers, configure Apache Kafka manually, replace servers when they fail, orchestrate server patches and upgrades, architect the cluster for high availability, ensure data is durably stored and secured, setup monitoring and alarms, and carefully plan scaling events to support load changes. Amazon Managed Streaming for Kafka makes it easy for you to build and run production applications on Apache Kafka without needing Apache Kafka infrastructure management expertise. That means you spend less time managing infrastructure and more time building applications.

With a few clicks in the Amazon MSK console you can create highly available Apache Kafka clusters with settings and configuration based on Apache Kafka's deployment best practices. Amazon MSK automatically provisions and runs your Apache Kafka clusters. Amazon MSK continuously monitors cluster health and automatically replaces unhealthy nodes with no downtime to your application. In addition, Amazon MSK secures your Apache Kafka cluster by encrypting data at rest.

CHAPTER 6

APPLICATION INTEGRATION

AWS Step Functions

AWS Step Functions lets you coordinate multiple AWS services into serverless workflows so you can build and update apps quickly. Using Step Functions, you can design and run workflows that stitch together services such as AWS Lambda and Amazon ECS into feature-rich applications. Workflows are made up of a series of steps, with the output of one step acting as input into the next. Application development is simpler and more intuitive using Step Functions, because it translates your workflow into a state machine diagram that is easy to understand, easy to explain to others, and easy to change. You can monitor each step of execution as it happens, which means you can identify and fix problems quickly. Step Functions automatically triggers and tracks each step, and retries when there are errors, so your application executes in order and as expected.

Amazon MQ

Amazon MQ is a managed message broker service for Apache ActiveMQ that makes it easy to set up and operate message brokers in the cloud. Message brokers allow different software systems–often using different programming languages, and on different platforms–to communicate and exchange information. Amazon MQ reduces your operational load by managing the provisioning, setup, and maintenance of ActiveMQ, a popular open-source message broker. Connecting your current applications to Amazon MQ is easy because it uses industry-standard APIs and protocols for messaging, including JMS, NMS, AMQP, STOMP, MQTT, and WebSocket. Using standards means that in most cases, there's no need to rewrite any messaging code when you migrate to AWS.

Amazon SQS

Amazon Simple Queue Service (Amazon SQS) is a fully managed message queuing service that enables you to decouple and scale microservices, distributed systems, and serverless applications. SQS eliminates the complexity and overhead associated with managing and operating message oriented middleware, and empowers developers to focus on differentiating work. Using SQS, you can send, store, and receive messages between software components at any volume, without losing messages or requiring other services to be available. Get started with SQS in minutes using the AWS console, Command Line Interface or SDK of your choice, and three simple commands.

SQS offers two types of message queues. Standard queues offer maximum throughput, best-effort ordering, and at-least-once delivery. SQS FIFO queues are designed to guarantee that messages are processed exactly once, in the exact order that they are sent.

Amazon SNS

Amazon Simple Notification Service (Amazon SNS) is a highly available, durable, secure, fully managed pub/sub messaging service that enables you to decouple microservices, distributed systems, and serverless applications. Amazon SNS provides topics for high-throughput, push-based, many-to-many messaging. Using Amazon SNS topics, your publisher systems can fan out messages to a large number of subscriber endpoints for parallel processing, including Amazon SQS queues, AWS Lambda functions, and HTTP/S webhooks. Additionally, SNS can be used to fan out notifications to end users using mobile push, SMS, and email.

Amazon SWF

Amazon Simple Workflow (Amazon SWF) helps developers build, run, and scale background jobs that have parallel or sequential steps. You can think of Amazon SWF as a fully-managed state tracker and task coordinator in the cloud. If your application's steps take more than 500 milliseconds to complete, you need to track the state of processing. If you need to recover or retry if a task fails, Amazon SWF can help you.

AR and VR

Amazon Sumerian

Amazon Sumerian lets you create and run virtual reality (VR), augmented reality (AR), and 3D applications quickly and easily without requiring any specialized programming or 3D graphics expertise. With Sumerian, you can build highly immersive and interactive scenes that run on popular hardware such as Oculus Go, Oculus Rift, HTC Vive, HTC Vive Pro, Google Daydream, and Lenovo Mirage as well as Android and iOS mobile devices. For example, you can build a virtual classroom that lets you train new employees around the world, or you can build a virtual environment that enables people to tour a building remotely. Sumerian makes it easy to create all the building blocks needed to build highly immersive and interactive 3D experiences including adding objects (e.g. characters, furniture, and landscape), and designing, animating, and scripting environments. Sumerian does not require specialized expertise and you can design scenes directly from your browser.

AWS Cost Management
AWS Cost Explorer
AWS Cost Explorer has an easy-to-use interface that lets you visualize, understand, and manage your AWS costs and usage over time. Get started quickly by creating custom reports (including charts and tabular data) that analyze cost and usage data, both at a high level (e.g., total costs and usage across all accounts) and for highly-specific requests (e.g., m2.2xlarge costs within account Y that are tagged "project: secretProject").

AWS Budgets
AWS Budgets gives you the ability to set custom budgets that alert you when your costs or usage exceed (or are forecasted to exceed) your budgeted amount. You can also use AWS Budgets to set RI utilization or coverage targets

and receive alerts when your utilization drops below the threshold you define. RI alerts support Amazon EC2, Amazon RDS, Amazon Redshift, and Amazon ElastiCache reservations.

Budgets can be tracked at the monthly, quarterly, or yearly level, and you can customize the start and end dates. You can further refine your budget to track costs associated with multiple dimensions, such as AWS service, linked account, tag, and others. Budget alerts can be sent via email and/or Amazon Simple Notification Service (SNS) topic.

Budgets can be created and tracked from the AWS Budgets dashboard or via the Budgets API.

AWS Cost & Usage Report

The AWS Cost & Usage Report is a single location for accessing comprehensive information about your AWS costs and usage.

The AWS Cost & Usage Report lists AWS usage for each service category used by an account and its IAM users in hourly or daily line items, as well as any tags that you have activated for cost allocation purposes. You can also customize the AWS Cost & Usage Report to aggregate your usage data to the daily or monthly level.

Reserved Instance (RI) Reporting

AWS provides a number of RI-specific cost management solutions out-of-the-box to help you better understand and manage your RIs. Using the RI Utilization and Coverage reports available in AWS Cost Explorer, you can visualize your RI data at an aggregate level or inspect a particular RI subscription. To access the most detailed RI information available, you can leverage the AWS Cost & Usage Report. You can also set a custom RI utilization target via AWS Budgets and receive alerts when your utilization drops below the threshold you define.

Blockchain

Amazon Managed Blockchain

Amazon Managed Blockchain is a fully managed service that makes it easy to create and manage scalable blockchain networks using the popular open source frameworks Hyperledger Fabric and Ethereum.

Blockchain makes it possible to build applications where multiple parties can execute transactions without the need for a trusted, central authority. Today, building a scalable blockchain network with existing technologies is complex to set up and hard to manage. To create a blockchain network, each network member needs to manually provision hardware, install software, create and manage certificates for access control, and configure networking components. Once the blockchain network is running, you need to continuously monitor the infrastructure and adapt to changes, such as an increase in transaction requests, or new members joining or leaving the network.

Amazon Managed Blockchain is a fully managed service that allows you to set up and manage a scalable blockchain network with just a few clicks. Amazon Managed Blockchain eliminates the overhead required to create the network, and automatically scales to meet the demands of thousands of applications running millions of transactions. Once your network is up and running, Managed Blockchain makes it easy to manage and maintain your blockchain network. It manages your certificates, lets you easily invite new members to join the network, and tracks operational metrics such as usage of compute, memory, and storage resources. In addition, Managed Blockchain can replicate an immutable copy of your blockchain network activity into Amazon Quantum Ledger Database (QLDB), a fully managed ledger database. This allows you to easily analyze the network activity outside the network and gain insights into trends.

Business Applications

Alexa for Business

Alexa for Business is a service that enables organizations and employees to use Alexa to get more work done. With Alexa for Business, employees can use Alexa as their intelligent assistant to be more productive in meeting rooms, at their desks, and even with the Alexa devices they already have at home.

Amazon WorkDocs

Amazon WorkDocs is a fully managed, secure enterprise storage and sharing service with strong administrative controls and feedback capabilities that improve user productivity.

Users can comment on files, send them to others for feedback, and upload new versions without having to resort to emailing multiple versions of their files as attachments.

Users can take advantage of these capabilities wherever they are, using the device of their choice, including PCs, Macs, tablets, and phones. Amazon WorkDocs offers IT administrators the option of integrating with existing corporate directories, flexible sharing policies and control of the location where data is stored. You can get started using Amazon WorkDocs with a 30-day free trial providing 1 TB of storage per user for up to 50 users.

Amazon WorkMail

Amazon WorkMail is a secure, managed business email and calendar service with support for existing desktop and mobile email client applications. Amazon WorkMail gives users the ability to seamlessly access their email, contacts, and calendars using the client application of their choice, including Microsoft Outlook, native iOS and Android email applications, any client application supporting the IMAP protocol, or directly through a web browser. You can integrate Amazon WorkMail with your existing corporate directory, use email journaling to meet compliance requirements, and control both the keys that encrypt your data and the location in which your data is stored. You can also set up interoperability with Microsoft Exchange Server, and programmatically manage users, groups, and resources using the Amazon WorkMail SDK.

Amazon Chime

Amazon Chime is a communications service that transforms online meetings with a secure, easy-to-use application that you can trust. Amazon Chime works seamlessly across your devices so that you can stay connected. You can use Amazon

Chime for online meetings, video conferencing, calls, chat, and to share content, both inside and outside your organization.

Amazon Chime works with Alexa for Business, which means you can use Alexa to start your meetings with your voice. Alexa can start your video meetings in large conference rooms, and automatically dial into online meetings in smaller huddle rooms and from your desk.

Database

Amazon Aurora

Amazon Aurora is a MySQL and PostgreSQL compatible relational database engine that combines the speed and availability of high-end commercial databases with the simplicity and cost-effectiveness of open source databases.

Amazon Aurora is up to five times faster than standard MySQL databases and three times faster than standard PostgreSQL databases. It provides the security, availability, and reliability of commercial databases at 1/10th the cost. Amazon Aurora is fully managed by Amazon Relational Database Service (RDS), which automates time-consuming administration tasks like hardware provisioning, database setup, patching, and backups.

Amazon Aurora features a distributed, fault-tolerant, self-healing storage system that auto-scales up to 64TB per database instance. It delivers high performance and availability with up to 15 low-latency read replicas, point-in-time recovery, continuous backup to Amazon S3, and replication across three Availability Zones (AZs).

Amazon RDS

Amazon Relational Database Service (Amazon RDS) makes it easy to set up, operate, and scale a relational database in the cloud. It provides cost-efficient and resizable capacity while automating time-consuming administration tasks such as hardware provisioning, database setup, patching and backups. It frees you to focus on your applications so you can give them the fast performance, high availability, security and compatibility they need.

Amazon RDS is available on several database instance types - optimized for memory, performance or I/O - and provides you with six familiar database engines to choose from, including Amazon Aurora, PostgreSQL, MySQL, MariaDB, Oracle Database, and SQL Server. You can use the AWS Database Migration Service to easily migrate or replicate your existing databases to Amazon RDS.

Amazon RDS on VMware

Amazon Relational Database Service (RDS) on VMware lets you deploy managed databases in on-premises VMware environments using the Amazon RDS technology enjoyed by hundreds of thousands of AWS customers. Amazon RDS provides cost-efficient and resizable capacity while automating time-consuming administration tasks including hardware provisioning, database setup, patching, and backups, freeing you to focus on your applications. RDS on VMware brings these same benefits to your on-premises deployments, making it easy to set up, operate, and scale databases in VMware vSphere private data centers, or to migrate them to AWS. RDS on VMware allows you to utilize the same simple interface for managing databases in on-premises VMware environments as you would use in AWS. You can easily replicate RDS on VMware databases to RDS instances in AWS, enabling low-cost hybrid deployments for disaster recovery, read replica bursting, and optional long-term backup retention in Amazon Simple Storage Service (Amazon S3).

Amazon DynamoDB

Amazon DynamoDB is a key-value and document database that delivers single-digit millisecond performance at any scale. It's a fully managed, multiregion, multimaster database with built-in security, backup and restore, and in-memory caching for internet-scale applications. DynamoDB can handle more than 10 trillion requests per day and support peaks of more than 20 million requests per second.

Many of the world's fastest growing businesses such as Lyft, Airbnb, and Redfin as well as enterprises such as Samsung, Toyota, and Capital One depend on the scale and performance of DynamoDB to support their mission-critical workloads.

More than 100,000 AWS customers have chosen DynamoDB as their key-value and document database for mobile, web, gaming, ad tech, IoT, and other applications that need low-latency data access at any scale. Create a new table for your application and let DynamoDB handle the rest.

Amazon ElastiCache

Amazon ElastiCache is a web service that makes it easy to deploy, operate, and scale an in-memory cache in the cloud.

The service improves the performance of web applications by allowing you to retrieve information from fast, managed, in-memory caches, instead of relying entirely on slower disk-based databases.

Amazon ElastiCache supports two open-source in-memory caching engines:

• • Redis - a fast, open source, in-memory data store and cache. Amazon ElastiCache for Redis is a Redis- compatible in-memory service that delivers the ease-of-use and power of Redis along with the availability, reliability, and performance suitable for the most demanding applications. Both single- node and up to 15-shard clusters are available, enabling scalability to up to 3.55 TiB of in-memory data. ElastiCache for Redis is fully managed, scalable, and secure. This makes it an ideal candidate to power high-performance use cases such as web, mobile apps, gaming, ad-tech, and IoT.

• • Memcached - a widely adopted memory object caching system. ElastiCache for Memcached is protocol compliant with Memcached, so popular tools that you use today with existing Memcached environments will work seamlessly with the service.

Amazon Neptune

Amazon Neptune is a fast, reliable, fully-managed graph database service that makes it easy to build and run applications that work with highly connected datasets. The core of Amazon Neptune is a purpose-built, high-performance graph database engine optimized for storing billions of relationships and querying the graph with milliseconds latency. Amazon Neptune supports popular graph models Property Graph and W3C's RDF, and their respective query languages Apache TinkerPop Gremlin and SPARQL, allowing you to easily build queries that efficiently navigate highly connected datasets. Neptune powers graph use cases such as recommendation engines, fraud detection, knowledge graphs, drug discovery, and network security.

Amazon Neptune is highly available, with read replicas, point-in-time recovery, continuous backup to Amazon S3, and replication across Availability Zones. Neptune is secure with support for encryption at rest. Neptune is fully-managed, so you no longer need to worry about database management tasks such as hardware provisioning, software patching, setup, configuration, or backups.

Amazon Quantum Ledger Database (QLDB)

Amazon QLDB is a fully managed ledger database that provides a transparent, immutable, and cryptographically verifiable transaction log owned by a central trusted authority. Amazon QLDB tracks each and every application data change and maintains a complete and verifiable history of changes over time.

Ledgers are typically used to record a history of economic and financial activity in an organization. Many organizations build applications with ledger-like functionality because they want to maintain an accurate history of their applications' data, for example, tracking the history of credits and debits in banking transactions, verifying the data lineage of an insurance claim, or tracing movement of an item in a supply chain network. Ledger applications are often implemented using custom audit tables or audit trails created in relational databases. However, building audit functionality with relational databases is time-consuming and prone to human error. It requires custom development, and since relational databases are not inherently immutable, any unintended changes to the data are hard to track and verify. Alternatively, blockchain frameworks, such as Hyperledger Fabric and Ethereum, can also be used as a ledger. However, this adds complexity as you need to set-up an entire blockchain network with multiple nodes, manage its infrastructure, and require the nodes to validate each transaction before it can be added to the ledger.

Amazon QLDB is a new class of database that eliminates the need to engage in the complex development effort of building your own ledger-like applications. With QLDB, your data's change history is immutable – it cannot be altered or deleted – and using cryptography, you can easily verify that there have been no unintended modifications to your application's data. QLDB uses an immutable transactional log, known as a journal, that tracks each application data change and maintains a complete and verifiable history of changes over time. QLDB is easy to use because it provides developers with a familiar SQL-like API, a flexible document data model, and full support for transactions. QLDB is also serverless, so it automatically scales to support the demands of your application. There are no servers to manage and no read or write limits to configure. With QLDB, you only pay for what you use.

Amazon Timestream

Amazon Timestream is a fast, scalable, fully managed time series database service for IoT and operational applications that makes it easy to store and analyze trillions of events per day at 1/10th the cost of relational databases. Driven by the rise of IoT devices, IT systems, and smart industrial machines, time-series data — data that measures how things change over time — is one of the fastest growing data types. Time-series data has specific characteristics such as typically arriving in time order form, data is append-only, and queries are always over a time interval. While relational databases can store this data, they are inefficient at processing this data as they lack optimizations such as storing and retrieving data by time intervals. Timestream is a purpose-built time series database that efficiently stores and processes this data by time intervals. With Timestream, you can easily store and analyze log data for DevOps, sensor data for IoT applications, and industrial telemetry data for equipment maintenance. As your data grows over time, Timestream's adaptive query processing engine understands its location and format, making your data simpler and faster to analyze. Timestream also automates rollups, retention, tiering, and compression of data, so you can manage your data at the lowest possible cost. Timestream is serverless, so there are no servers to manage. It manages time-consuming tasks such as server provisioning, software patching, setup, configuration, or data retention and tiering, freeing you to focus on building your applications.

AMAZON WEB SERVICES

Desktop and App Streaming
Amazon WorkSpaces
Amazon WorkSpaces is a fully managed, secure cloud desktop service. You can use Amazon WorkSpaces to provision either Windows or Linux desktops in just a few minutes and quickly scale to provide thousands of desktops to workers across the globe. You can pay either monthly or hourly, just for the WorkSpaces you launch, which helps you save money when compared to traditional desktops and on-premises VDI solutions. Amazon WorkSpaces helps you eliminate the complexity in managing hardware inventory, OS versions and patches, and Virtual Desktop Infrastructure (VDI), which helps simplify your desktop delivery strategy. With Amazon WorkSpaces, your users get a fast, responsive desktop of their choice that they can access anywhere, anytime, from any supported device.

Amazon AppStream 2.0
Amazon AppStream 2.0 is a fully managed application streaming service. You centrally manage your desktop applications on AppStream 2.0 and securely deliver them to any computer. You can easily scale to any number of users across the globe without acquiring, provisioning, and operating hardware or infrastructure. AppStream 2.0 is built on AWS, so you benefit from a data center and network architecture designed for the most security-sensitive organizations. Each user has a fluid and responsive experience with your applications, including GPU-intensive 3D design and engineering ones, because your applications run on virtual machines (VMs) optimized for specific use cases and each streaming session automatically adjusts to network conditions.

Enterprises can use AppStream 2.0 to simplify application delivery and complete their migration to the cloud. Educational institutions can provide every student access to the applications they need for class on any computer. Software vendors can use AppStream 2.0 to deliver trials, demos, and training for their applications with no downloads or installations. They can also develop a full software-as-a-service (SaaS) solution without rewriting their application.

Lightning Source UK Ltd.
Milton Keynes UK
UKHW020954070521
383312UK00013B/945